ENERGY SCIENCE, ENGINEERING AND TECHNOLOGY

THE TRANSFORMATION OF ELECTRICAL POWER

KEY ISSUES

ENERGY SCIENCE, ENGINEERING AND TECHNOLOGY

Additional books in this series can be found on Nova's website under the Series tab.

Additional E-books in this series can be found on Nova's website under the E-book tab.

ENERGY POLICIES, POLITICS AND PRICES

Additional books in this series can be found on Nova's website under the Series tab.

Additional E-books in this series can be found on Nova's website under the E-book tab.

ENERGY SCIENCE, ENGINEERING AND TECHNOLOGY

THE TRANSFORMATION OF ELECTRICAL POWER

KEY ISSUES

DAWSON B. MUNOZ
EDITOR

nova publishers
New York

Library of Congress Cataloging-in-Publication Data

ISBN: 978-1-62618-619-4

Published by Nova Science Publishers, Inc. † New York

Contents

PREFACE

This book examines the electric power industry which is in a transformation process. The electricity infrastructure of the United States is aging, and uncertainty exists around how to modernize the grid, and what technologies and fuels will be used to produce electricity in the future. Unresolved questions of transmission and reliability of the grid are arising from potential cybersecurity threats and continuing interest in harnessing renewable energy and other low carbon sources of electricity. Concerns about reliability and electricity prices are being complicated by new environmental regulations and the rising availability of natural gas for the production of electric power from unconventional resources such as gas shales. Congress will likely be faced with policy issues regarding how the modernization of this vital industry will unfold.

Chapter 1 - The electric power industry is in the process of transformation. The electricity infrastructure of the United States is aging, and uncertainty exists around how to modernize the grid, and what technologies and fuels will be used to produce electricity in the future. Congress will likely be faced with policy issues regarding how the modernization of this vital industry will unfold. For most of the 20th century, coal has been the dominant fuel used to produce electricity. In 2011, coal was the fuel used for almost 42% of power generation in the United States. However, coal use for power generation seems to be on the decline. In April 2012, for the first time in history, the amount of electricity generated from natural gas equaled that of coal (according to Energy Information Administration statistics) with each fuel claiming about 32% of the market. The future of coal as a fuel for power generation seems to be in question. Two major reasons are generally seen as being responsible: the expectation of a dramatic rise in natural gas supplies, and the impact of environmental regulations on an aging base of coal-fired power plants. The electric utility industry values diversity in fuel choice options since reliance on one fuel or technology can leave electricity producers vulnerable to price and supply volatility. However, an "inverse relationship" is developing for coal vs. natural gas as a power generation choice based on market economics alone, and policies which allow one fuel source to dominate may come at the detriment of the other.

Upgrading the nation's transmission system to accommodate current and future uses, and ensuring the reliable functioning and the security of the grid, has been a major concern for the federal government. Federal law has already tasked the Federal Energy Regulatory Commission with responsibility for enforcing reliability standards for the bulk electric system, including cybersecurity, but protection from natural hazards continues as a key issue. The recent damage sustained to the electrical grid by Hurricane Sandy in New York and New Jersey and difficulty in restoring electricity service underscore the age and fragility of the power system, and how electricity service might benefit from hardening and modernization of various power systems. Growing concerns over greenhouse gas (GHG) emissions, other environmental costs associated with burning fossil fuels, and existing or anticipated state and federal policies addressing these issues are leading some utilities and energy providers to deploy more renewable energy technologies to meet power demands, and potentially increasing the need for new transmission lines to incorporate clean energy sources.

New environmental regulations under development would impose new requirements on coal-fired power plants. Some of these rules would be implemented at the federal level, while others would be implemented at the state level. The Environmental Protection Agency (EPA) also issued standards for greenhouse gas emissions which would require all new power plants to restrict carbon dioxide emissions. EPA has yet to propose rules for GHG emissions from existing power plants, as is required by court order. Much attention has focused on the resulting finalization of these regulations, and their potential to contribute to power plant retirements, with some in the electric power industry expressing concern that reliability could be impacted.

Chapter 2 - For most of the twentieth century, the primary use of coal in the United States was for electric power generation, and for most of the history of power generation in the United States, coal has been the dominant fuel used to produce electricity. Even as recently as 2011, coal was the fuel used for almost 42% of power generation in the United States accounting for 93% of coal use. Industrial uses represented the remaining 7%. However, in April 2012, coal's share of the power generation market dropped to about 32% (according to Energy Information Administration statistics), equal to that of natural gas. Coal was the fuel of choice because of its availability and the relatively low cost of producing electricity in large, coal-burning power plants which took advantage of coal's low-priced, high energy content to employ economies of scale in steam-electric production. However, coal use for power generation seems to be on the decline, and the magnitude of coal's role for power generation is in question. Two major reasons are generally seen as being responsible: the expectation of a dramatic rise in natural gas supplies, and the impact of environmental regulations on an aging base of coal-fired power plants.

A recent drop in natural gas prices has been enabled by increasing supplies of natural gas largely due to horizontal drilling and hydraulic fracturing (i.e., fracking) of shale gas formations. If the production can be sustained in an environmentally acceptable manner, then a long-term, relatively inexpensive supply of natural gas could result. Decreased natural gas prices are lowering wholesale electricity prices, stimulating a major switch from coal to gas-burning facilities. The electric utility industry values diversity in fuel choice options since reliance on one fuel or technology can leave electricity producers vulnerable to price and supply volatility. However, an "inverse relationship" may be developing for coal vs. natural gas as a power generation choice based on market economics alone, and policies which allow one fuel source to dominate may come at the detriment of the other.

Coal-fired power plants are among the largest sources of air pollution in the United States. More than half a dozen separate Clean Air Act programs could possibly be used to control emissions, which makes compliance strategy potentially complicated for utilities and difficult for regulators. Because the cost of the most stringent available controls, for the entire industry, could range into the tens of billions of dollars, some power companies have fought hard and rather successfully to limit or delay regulations affecting them, particularly with respect to plants constructed before the Clean Air Act Amendments of 1970 were passed.

The expected retirement of approximately 27 GW of coal-fired capacity by 2016 has been reported to the Energy Information Administration (EIA) by coal plant owners and operators, accounting for approximately 8.5% of U.S. coal-fired capacity. While the costs of compliance with new Environmental Protection Agency regulations are a factor, several other issues are cited by coal plant owners and operators as contributing to these retirement decisions including the age of coal-fired power plants, flat to modest electricity demand growth, the availability of previously underutilized natural gas combined-cycle power plants, and the lower price of natural gas due to shale gas development. Even coal plants which have made significant modifications to meet existing EPA regulations are being closed or mothballed due to a combination of low natural gas prices, and the inability to sell power into other markets. EIA expects coal to be a significant part of the U.S. power generation industry's future to well past 2030. But given price competition from natural gas, and emerging environmental regulations, that role will likely be smaller than in recent decades. Coal-fired generation is likely to face a challenging future.

Chapter 3 - The United States faces important decisions about future energy supply and use. A key question is how renewable energy resources might be used to meet U.S. energy needs in general, and to meet U.S. electricity needs specifically. Renewable energy sources are typically used for three general types of applications: electricity generation, biofuels/bioproducts, and heating/cooling. Each application uses different technologies to convert renewable energy sources into usable products. The literature on renewable energy resources, conversion technologies for different applications, and economics is massive. This report focuses on electricity generation from renewable energy sources. In 2010, renewable sources of energy were used to produce almost 11% (7% from hydropower and 4% from other renewables) of the 4 million gigawatthours of electricity generated in the United States.

This report provides a summary of U.S. electricity generation potential from wind, solar, geothermal, hydroelectric, ocean-hydrokinetic, and biomass sources of renewable energy. The focus of this report is twofold: (1) provide an assessment of U.S. renewable electricity generation potential and how renewables might satisfy electric power sector demand, and (2) discuss challenges, issues, and barriers that might limit renewable electricity generation deployment. Data sources from 15 different organizations were reviewed to derive estimates of electricity generation potential. One key finding is that there exists no uniform national assessment of renewable electricity generation potential. No standard methods or set of assumptions are used to estimate renewable electricity generation potential. So even existing assessments for individual energy sources are difficult to compare objectively. In order to compare various estimates on an equivalent basis, CRS engaged experts in each renewable energy resource area to help normalize electricity generation potential estimates into a common metric: gigawatthours per year.

Chapter 4 - Electricity is vital to the commerce and daily functioning of United States. The modernization of the grid to accommodate today's uses is leading to the incorporation of information processing capabilities for power system controls and operations monitoring. The "Smart Grid" is the name given to the evolving electric power network as new information technology systems and capabilities are incorporated. While these new components may add to the ability to control power flows and enhance the efficiency of grid operations, they also potentially increase the susceptibility of the grid to cyber (i.e., computer-related) attack since they are built around microprocessor devices whose basic functions are controlled by software programming. The potential for a major disruption or widespread damage to the nation's power system from a large scale cyberattack has increased focus on the cybersecurity of the Smart Grid. Federal efforts to enhance the cybersecurity of the electrical grid were emphasized with the recognition of cybersecurity as a critical issue for electric utilities in developing the Smart Grid. The Federal Energy Regulatory Commission (FERC) received primary responsibility for the reliability of the bulk power system from the Energy Policy Act of 2005. FERC subsequently designated the North American Electric Reliability Corporation (NERC) as the "Electric Reliability Organization" (ERO) with the responsibility of establishing and enforcing reliability standards. Compliance with reliability standards for electric utilities thus changed from a voluntary, peer-driven undertaking to a mandatory function. The Energy Independence and Security Act of 2007 (EISA) later added requirements for "a reliable and secure electricity infrastructure" with regard to Smart Grid development. NERC is also responsible for standards for critical infrastructure protection (CIP) which focus on planning and procedures for the physical security of the grid. Self-determination is a key part of the CIP reliability process. Utilities are allowed to self-identify what they see as "critical assets" under NERC regulations. Only "critical cyber assets" (i.e., as essential to the reliable operation of critical assets) are subject to CIP standards. FERC has directed NERC to revise the standards so that some oversight of the identification process for critical cyber assets was provided, but any revision is again subject to stakeholder approval. While reliability standards are mandatory, the ERO process for developing regulations is somewhat unusual in that the regulations are essentially being established by the entities who are being regulated. This may potentially be a conflict of interest, especially when cost of compliance is a concern, and acceptable standards may conceivably result from the option with the lowest costs. Since utility systems are interconnected in many ways, the system with the least protected network potentially provides the weakest point of access.

Cybersecurity threats represent a constantly moving and increasing target for mitigation activities and mitigation efforts could likewise spiral upward in costs. Recovery of costs may present a major challenge especially to distribution utilities and state commissions charged with overseeing utility costs. EISA only requires states to consider recovery of costs related to Smart Grid systems. FERC has jurisdiction over the bulk power grid, and cannot compel entities involved in distribution to comply with its regulations. Recoverability from a cyber attack on the scale of something which could take down a significant portion of the grid will likely be very difficult, but maintaining a ready inventory of critical spare parts in close proximity to key installations could quicken recovery efforts from some types of attack. The electricity grid is connected to (and largely dependent on) the natural gas pipeline, water supply, and telecommunications systems. Technologies being developed for use by the Smart Grid could also be used by these industries. Consideration could be given to applying similar control system device and system safeguards to these other critical utility systems.

Chapter 5 - High winds, especially when combined with precipitation from seasonal storms, can cause damage to electricity utility systems, resulting in service interruptions to large numbers of electricity customers. While most such power outages are caused by damage from trees and tree limbs falling on local electricity distribution lines and poles, major power outages tend to be caused by damage to electricity transmission lines which carry bulk power long distances. Depending on the severity of the storm and resulting impairment, power outages can last a few hours or extend to periods of several days, and have real economic effects. Power outages can impact businesses (primarily through lost orders and damage to perishable goods and inventories), and manufacturers (mainly through downtime and lost production, or equipment damage). Data from various studies lead to cost estimates from storm-related outages to the U.S. economy at between $20 billion and $55 billion annually. Data also suggest the trend of outages from weather-related events is increasing.

Suggested solutions for reducing impacts from weather-related outages include improved tree-trimming schedules to keep rights-of-way clear, placing distribution and some transmission lines underground, implementing Smart Grid improvements to enhance power system operations and control, inclusion of more distributed generation, and changing utility maintenance practices and metrics to focus on power system reliability. However, most of these potential solutions come with high costs which must be balanced against the perceived benefits. A number of options exist for Congress to consider which could help reduce storm-related outages. These range from improving the quality of data on storm-related outages, to a greater strategic investment in the U.S. electricity grid. Congress could empower a federal agency to develop standards for the consistent reporting of power outage data. While responsibility for the reliability of the bulk electric system is under the Federal Energy Regulatory Commission (as per the Energy Policy Act of 2005), no central responsibility exists for the reliability of distribution systems. One possible option could be to bring distribution systems under the Electric Reliability Organization for reliability purposes. Recovery after storm-related outages might be enhanced by a federal role in formalizing the review or coordination of electric utility mutual assistance agreements (MAAs). This would not necessarily mean federal approval of MAAs, but may help in the cooperative coordination of additional federal and state resources, especially in a wide, multi-state weather event. While there has been much discussion of transmission system inadequacies and inefficiencies, many distribution systems are in dire need of upgrades or repairs. The cost of upgrading the U.S. grid to meet future uses is expected to be high, with the American Society of Civil Engineers estimating a need of $673 billion by 2020. While the federal government recently made funding available of almost $16 billion for specific Smart Grid projects and new transmission lines under the American Recovery and Reinvestment Act of 2009, there has not been a comprehensive effort to study the needs, set goals, and provide targeted funding for modernization of the U.S. grid as part of a long-term national energy strategy. Such an effort would also require decisions about the appropriate roles of government and the private sector. Power delivery systems are most vulnerable to storms and extreme weather events. Improving the overall condition and efficiency of the power delivery system can only serve to improve the resiliency of the system, and help hasten recovery from weather-related outages. Ultimately, however, electric utilities are responsible for this infrastructure. They are in the business of selling electricity, and they cannot sell electricity if their power delivery systems are out of service.

Chapter 6 - On December 21, 2011, EPA Administrator Lisa Jackson announced final standards aimed at reducing mercury and other air toxics emissions from electric generating units (EGUs) by about 90%. The rule, commonly referred to as the "Utility MACT" or the "Mercury and Air Toxics Standards" (MATS), has been more than a decade in the making (Congress authorized the standards in the 1990 Clean Air Act Amendments), and it is among the most expensive rules that EPA has ever promulgated. EPA estimates the annualized cost at $9.6 billion in 2015. Industry estimates have been higher. The benefits are also large, according to EPA, ranging from $37 billion to $90 billion annually. The benefits mostly reflect the monetized value of avoiding up to 11,000 premature deaths annually. The rule's costs will fall primarily on older coal-fired units that do not have state-of-the art pollution controls. EPA says that this is a minority of coal-fired plants and an even smaller share of all electric generation: the agency estimates that 56% of coal-fired units have already installed equipment that can be used to meet the standards. In addition, about 55% of the nation's electricity supply comes from natural gas, nuclear, and renewable sources that are not subject to the rule's requirements. This report describes the rule and its potential impact. The report discusses previous EPA efforts to regulate utility mercury emissions, the court decision overturning those regulations, the specifics of the new rule, its estimated costs and benefits, the impact of the rule on electric.

Chapter 7 - The policy debate over the role of nuclear power in the nation's energy mix is rooted in the technology's fundamental characteristics. Nuclear reactors can produce potentially vast amounts of energy with relatively low consumption of natural resources and emissions of greenhouse gases and other pollutants. However, facilities that produce nuclear fuel for civilian power reactors can also produce materials for nuclear weapons. The process of nuclear fission (splitting of atomic nuclei) to generate power also results in the production of radioactive material that must be contained in the reactor and can remain hazardous for thousands of years. How to manage the weapons proliferation and safety risks of nuclear power, or whether nuclear power is worth those risks, are issues that have long been debated in Congress. The 104 licensed nuclear power reactors at 65 sites in the United States generate about 20% of the nation's electricity. Five new reactors are currently licensed for construction. About a dozen more are planned, but whether they move forward will depend largely on their economic competitiveness with natural gas and coal plants. Throughout the world, 436 reactors are currently in service, and 62 more are under construction. The March 2011 disaster at the Fukushima Dai-ichi nuclear power plant in Japan increased attention to nuclear safety throughout the world. The U.S. Nuclear Regulatory Commission (NRC), which issues and enforces nuclear safety requirements, established a task force to identify lessons from Fukushima applicable to U.S. reactors. The task force's report led to NRC's first Fukushima-related regulatory requirements on March 12, 2012. Several other countries, such as Germany and Japan, eliminated or reduced their planned future reliance on nuclear power after the accident. Highly radioactive spent nuclear fuel that is regularly removed from nuclear power plants is currently stored at plant sites in the United States. Plans for a permanent underground repository at Yucca Mountain, NV, were abandoned by the Obama Administration, although that decision is being challenged in court. The Obama Administration appointed the Blue Ribbon Commission on America's Nuclear Future to recommend an alternative nuclear waste policy. The Commission recommended in January 2012 that new candidate sites for nuclear waste storage and disposal facilities be selected through a "consent based" process. The level of security that must be provided at nuclear power plants has been a high-profile issue since the 9/11 terrorist attacks on the United States in 2001. Since those attacks, NRC issued a series of orders and regulations that substantially increased nuclear plant security requirements, although industry critics contend that those measures are still insufficient. Encouraging exports of U.S. civilian nuclear products, services, and technology while making sure they are not used for foreign nuclear weapons programs has long been a fundamental goal of U.S. nuclear energy policy. Recent proposals to build nuclear power plants in several countries in the less developed world, including the Middle East, have prompted concerns that international controls may prove inadequate.

Chapter 8 – The report by U.S. Energy Information Administration on Fuel Used in Electricity Generation, dated July 30, 2012.

In: The Transformation of Electrical Power: Key Issues
Editor: Dawson B. Munoz
ISBN: 978-1-62618-619-4

Chapter 1

ELECTRICAL POWER: OVERVIEW OF CONGRESSIONAL ISSUES*

Richard J. Campbell

SUMMARY

The electric power industry is in the process of transformation. The electricity infrastructure of the United States is aging, and uncertainty exists around how to modernize the grid, and what technologies and fuels will be used to produce electricity in the future. Congress will likely be faced with policy issues regarding how the modernization of this vital industry will unfold.

For most of the 20th century, coal has been the dominant fuel used to produce electricity. In 2011, coal was the fuel used for almost 42% of power generation in the United States. However, coal use for power generation seems to be on the decline. In April 2012, for the first time in history, the amount of electricity generated from natural gas equaled that of coal (according to Energy Information Administration statistics) with each fuel claiming about 32% of the market. The future of coal as a fuel for power generation seems to be in question. Two major reasons are generally seen as being responsible: the expectation of a dramatic rise in natural gas supplies, and the impact of environmental regulations on an aging base of coal-fired power plants. The electric utility industry values diversity in fuel choice options since reliance on one fuel or technology can leave electricity producers vulnerable to price and supply volatility. However, an "inverse relationship" is developing for coal vs. natural gas as a power generation choice based on market economics alone, and policies which allow one fuel source to dominate may come at the detriment of the other.

Upgrading the nation's transmission system to accommodate current and future uses, and ensuring the reliable functioning and the security of the grid, has been a major concern for the federal government. Federal law has already tasked the Federal Energy Regulatory Commission with responsibility for enforcing reliability standards for the bulk electric

* This is an edited, reformatted and augmented version of the Congressional Research Service Publication, CRS Report for Congress R42923, dated January 30, 2013.

system, including cybersecurity, but protection from natural hazards continues as a key issue. The recent damage sustained to the electrical grid by Hurricane Sandy in New York and New Jersey and difficulty in restoring electricity service underscore the age and fragility of the power system, and how electricity service might benefit from hardening and modernization of various power systems. Growing concerns over greenhouse gas (GHG) emissions, other environmental costs associated with burning fossil fuels, and existing or anticipated state and federal policies addressing these issues are leading some utilities and energy providers to deploy more renewable energy technologies to meet power demands, and potentially increasing the need for new transmission lines to incorporate clean energy sources.

New environmental regulations under development would impose new requirements on coal-fired power plants. Some of these rules would be implemented at the federal level, while others would be implemented at the state level. The Environmental Protection Agency (EPA) also issued standards for greenhouse gas emissions which would require all new power plants to restrict carbon dioxide emissions. EPA has yet to propose rules for GHG emissions from existing power plants, as is required by court order. Much attention has focused on the resulting finalization of these regulations, and their potential to contribute to power plant retirements, with some in the electric power industry expressing concern that reliability could be impacted.

SYNTHESIS OF KEY ISSUES

The electric power industry is in the process of transformation. The electricity infrastructure of the United States is aging, and uncertainty exists around how to modernize the grid, and what technologies and fuels will be used to produce electricity in the future. Unresolved questions of transmission and reliability of the grid are arising from potential cybersecurity threats and continuing interest in harnessing renewable energy and other low carbon sources of electricity. Concerns about reliability and electricity prices are being complicated by new environmental regulations and the rising availability of natural gas for the production of electric power from unconventional resources such as gas shales. Congress will likely be faced with policy issues regarding how the modernization of this vital industry will unfold.

BASIC FACTS AND STATISTICS

Electric Power Generation in the United States

The electrical grid of the United States consists of all the power plants generating electricity, together with the transmission and distribution lines and their associated transformers and substations which bring power to end-use customers. Electric power generation in the United States is currently dominated by the use of combustible fuels, such as the fossil fuels coal and natural gas, or from biomass. These fuels are burned either to produce steam in boilers which is used to turn turbine-generators or burned directly in combustion turbines which turn generators to produce electricity. Nuclear power uses heat from the fission of radioactive elements to produce steam. However, electricity can also be generated directly by wind turbines, solar power, geothermal energy, and hydropower. Generally, electricity must be used as soon as it is produced, because large amounts of electricity cannot be easily stored.

Originally, the individual utility company systems were not linked, but with greater electricity demand came the necessity of sharing generation resources. This sharing of generation resources required an interconnection of separate company systems to enable power sales and transfers.

These aggregated power systems form three major "interconnections"—the Eastern and Western interconnections, and the Electric Reliability Council of Texas, which includes most of that state. Within these interconnections are reliability regions, and a number of balancing authorities[1] which "integrate resource plans ahead of time, maintain load-interchange generation balance" within a balancing authority area, and "support interconnection frequency in real-time." The grid also connects the many publicly and privately owned electric utility and power companies in different states and regions of the United States[2] (and in Canada and Mexico).

Congress passed the Public Utility Regulatory Policies Act of 1978 (P.L. 95-617), creating a class of non-utility power producers (i.e., qualifying small power and cogeneration facilities) which resulted in the introduction of competition into wholesale power markets (e.g., for the sale of electricity to entities other than the end-user of power).

The passage of the Energy Policy Act of 1992 (P.L. 102-486) served to further promote greater competition in the bulk power markets.[3] As a result, in many parts of the United States, the electric power industry began to transition from highly regulated, local monopoly companies which generated, transmitted, and distributed electricity to end-use customers, to a business in which power generation is competitive while the industry's transmission and distribution functions are still highly regulated.

In these regions with restructured electricity industries, competitive markets largely set the price of power. The rates consumers ultimately pay for electricity are based on auctions in regional transmission organization (RTO) or independent system operator (ISO) systems wherein generators competitively bid to provide energy for a particular time period. Rates for wholesale transactions of RTOs and ISOs are under the Federal Energy Regulatory Commission's (FERC's) regulatory jurisdiction. Wholesale power rate components depend upon the market structure of the RTO or ISO, and generally have cost components for energy, capacity, transmission, ancillary services,[4] operating reserves, and general system costs and regulatory fees. Non-RTO transmission costs and distribution costs are generally added to arrive at the rate end-use customers pay. While FERC is largely responsible for regulation of the electric power transmission system and wholesale power markets, regulation of the distribution function of the electric power business is still largely carried out by state government agencies.

The choice of power generation technology in the United States is heavily influenced by the cost of fuel. Historically, the use of fossil fuels has provided some of the lowest prices for generating electricity.

Figure 1 shows that, as of 2011, coal accounts for approximately 42% of net generation by the electric power sector,[5] followed by natural gas at 25%, and nuclear power at almost 20%.

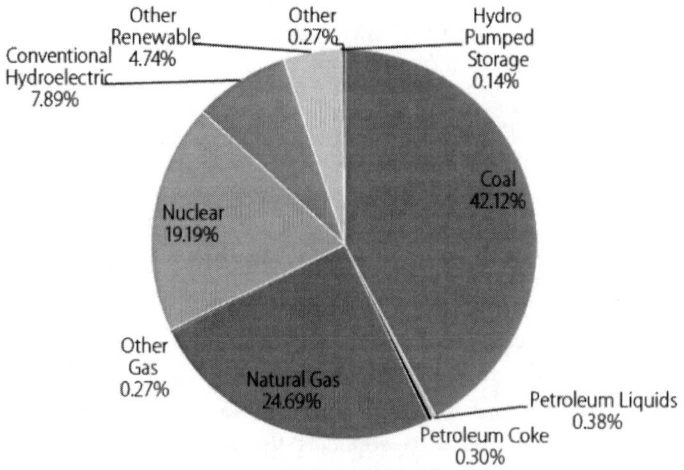

Source: U.S. Energy Information Administration, Annual Energy Review 2011. Table 8.2a.

Notes: "Other" includes batteries, chemicals, hydrogen, pitch, purchased steam, sulfur, miscellaneous technologies, and non-renewable waste (municipal solid waste from non-biogenic sources, and tire-derived fuels). See http://www.eia.gov/emeu/ aer/elect.html.

Figure 1. U.S. Electric Power Industry Net Generation, 2011.

MAJOR ELECTRIC POWER ISSUES

Future Technologies for Electricity Production Face Uncertainty

Electricity generation is vital to the commerce and daily functioning of the United States. However, the average age of power plants is now over 30 years, and the life expectancy of most power plants is about 40 years. While most of these plants are well-maintained, they are generally not as efficient as newer power plants. As power plants age, they are generally upgraded to continue operations, but the least efficient plants may be retired. Other plants may be shifted from base load operations (in which they essentially operate around the clock) to less demanding intermediate or peaking schedules. The cost of building a power plant is generally recovered over the depreciable life of the asset, such that operations and maintenance (O&M) expenses become the major component of an older power plant's continuing costs. A major component of O&M is the cost of fuel. However, the costs of modernizing older plants to meet new regulatory requirements can be relatively high. When the cost of upgrades to meet new environmental requirements is considered along with (perhaps increasing) O&M expenses, many older power plants become subject to retirement decisions.

Electric power generation is responsible for 37% of U.S. domestic carbon dioxide emissions (the primary anthropogenic greenhouse gas (GHG)), and over one-third of all U.S. GHG emissions.[6]

Growing concerns over GHG emissions, other environmental costs associated with burning fossil fuels, and existing or anticipated state and federal policies addressing these issues are leading some utilities and energy providers to deploy more renewable energy technologies to meet power demands. As of 2011, hydropower represented 8% of all U.S. electric power industry net generation, with all other renewable energy accounting for a further combined total of 4%.[7]

Renewable energy has been used since before the Industrial Revolution, but not on the scale of steam power generation. Renewable energy technologies use the power of the sun, wind, water, and heat from the earth, offering the possibility of producing electricity on a large scale without most of the environmental and climate consequences of electric power generation using fossil fuels.

Renewable energy sources have the potential to provide inexpensive, almost limitless electricity with minimal adverse environmental impacts. However, some of the technologies used today to generate electricity from renewable energy sources are variable in nature, and produce higher cost power than conventional fossil or nuclear sources of electricity (if environmental externalities are not considered). Practical energy storage technologies can potentially be a game-changer for greater renewable energy deployment. State governments have generally led the way in encouraging deployment of renewable energy technologies, using a Renewable Portfolio Standard (RPS) to create a market for renewable energy via mandated requirements. While most RPS goals are expected to be met, about 12 states have existing provisions expiring by 2015, and approximately 14 states and the District of Columbia have existing RPS or related provisions scheduled to expire by 2020.

Recent Events

Electricity today is widely viewed as a commodity.[8] As a commodity, electricity is primarily bought and sold as both power[9] and energy,[10] with various attributes being traded in electricity markets.

Competitive electricity markets have enabled a variety of wholesale electricity products and services to facilitate the sale and transmission of power. These involve both physical transactions (i.e., electricity is generated and sent to or taken off the grid), and financial transactions (i.e., the purchase and sale of electricity). Services have also arisen to provide transaction flexibility, and to manage (or hedge) the risks of various transactions. Some purchasers of electricity as a commodity do so solely for financial reasons. With the California (or Western) energy crisis of 2000 to 2001, the susceptibility of electricity markets to manipulation became evident.[11] Enron and its affiliates were principally found liable for "engaging in various gaming and market manipulation schemes," with an initial decision ordering the disgorgement of $1.6 billion in unjust profits.[12] FERC continues to investigate allegations of energy market manipulation, with nine actions in FY2012 resulting in civil penalties of $148 million assessed, and disgorgement of unjust profits of another $121 million.[13]

Increases in the domestic production of natural gas (primarily due to hydraulic fracturing of gas shales) are causing a dramatic change in electric power production decisions. With increasing production of natural gas has come a decline in natural gas prices, and a decline in coal consumption for power generation. In April 2012, for the first time in history, the amount of electricity generation from natural gas equaled that of coal, according to EIA statistics, with each representing about 32% of the market.

Many in the energy industry believe that a structural change in the economics of natural gas use has begun.

> While coal is projected to retain the largest share of the electricity generation mix through 2035, analyses included in the Annual Energy Outlook 2012 ... anticipate its share declining as more generation comes from natural gas and renewable technologies. Coal's role as the preeminent source of electricity generation in the United States has lessened in recent years, declining from 49% of total electricity generation in 2007 to 42% in 2011.... Projected fuel prices and economic growth are key factors influencing the future electricity generation mix. The price of natural gas, coal's chief competitor, has dropped significantly in recent years due to the increase in domestic production of natural gas.[14]

The electric utility industry values diversity in fuel choice options since reliance on one fuel or technology can leave electricity producers vulnerable to price and supply volatility. However, an "inverse relationship" is developing for coal vs. natural gas as a power generation choice based on market economics alone, and policies which allow one fuel source to dominate could increase industry vulnerability to volatility. Nevertheless, according to the EIA, coal is expected to be a key part of electricity generation in the United States well past the year 2030.[15] Increased availability of natural gas also raises the prospect of a greater use of distributed generation for power generation.

Selected Congressional Action from 112th Congress

Clean Energy Standard Act of 2012 (S. 2146, Bingaman)

Would have amended the Public Utility Regulatory Policies Act of 1978 to require, beginning in calendar 2015, each electric utility that sold electric energy to electric consumers in a state (other than Alaska or Hawaii) to obtain a percentage of such electric energy from clean energy. This minimum requirement would have risen from 24% in 2015, to a minimum of 84% in 2035 and the years thereafter. The bill directed the Secretary of Energy to (1) establish a federal clean energy credit trading program under which electric utilities may submit clean energy credits to certify their compliance, and (2) issue to each generator of electric energy a quantity of clean energy credits determined in accordance with the bill. Any electric utility that failed to meet the requirements of the bill would have been subject to a civil penalty. The bill would have required the Secretary to establish a state energy efficiency funding program. Introduced March 1, 2012; hearings held by Committee on Energy and Natural Resources.

American Renewable Energy and Efficiency Act (H.R. 5967, Markey)

Would have amended the Public Utility Regulatory Policies Act of 1978 to require the Federal Energy Regulatory Commission to promulgate regulations to implement and enforce a federal combined efficiency and renewable electricity standard. Would have required each retail electric supplier to submit to FERC, annually, an amount of federal renewable electricity credits and demonstrated total annual electricity savings for the previous year that, in the aggregate, was equal to the supplier's annual combined target for that year as established by the bill (compliance obligation). The bill would have required FERC, upon a request from a state's governor, to increase, to no more than half, the proportion of the annual combined targets for suppliers located within such state that could be met through submission of electricity savings (as opposed to renewable electricity credits). Would have required a supplier's target to be equal to the product of the supplier's base amount (electricity sold) for the year and a specified annual percentage for that year, which increases from 8% for 2014 to 50% for 2035 through 2040. The bill would have prescribed penalties for noncompliance with, or violations of, this bill. The bill would terminate on December 31, 2041. The bill would have authorized states to set the rates for a sale of electricity by a facility generating electricity from renewable energy sources pursuant to a state-approved production incentive program. Introduced June 19, 2012; Referred to the House Committee on Energy and Commerce.

Additional Reference

Renewable Electricity Futures Study, National Renewable Energy Laboratory, 2012, http://www.nrel.gov/docs/fy12osti/52409-1.pdf.

ADEQUACY AND RELIABILITY OF THE GRID

As with electric power plants, electric transmission and distribution system components are also aging, with power transformers averaging over 40 years of age, and 70% of transmission lines being 25 years or older. As components of the system are retired, they are replaced with newer components often linked to communications or automated systems (i.e., the Smart Grid). With changes in federal law encouraging electricity competition and markets, regulatory changes, and the aging of the electric power infrastructure as drivers, the grid is changing from a largely patchwork system built to serve the needs of individual electric utility companies to essentially a national interconnected system capable of accommodating massive transfers of electrical energy between regions of the United States.

Damage from storms to transmission and distribution systems appears to be increasing. But undergrounding transmission and distribution systems is not always a solution. Underground lines can be more expensive, and take longer to repair. Replacing overhead lines with underground cable is also expensive.

Upgrading the nation's transmission to accommodate current and future uses and ensuring the functioning and the security of the grid have been paramount concerns for the federal government. The federal government has already tasked the FERC with responsibility for enforcing reliability standards for the bulk electric system, but cybersecurity[16] and physical security of the grid remain key issues. The information technology systems and capabilities of the Smart Grid add to the ability to control power flows and enhance the efficiency of grid operations. But these attributes also potentially increase the susceptibility of the grid to cyber (i.e., computer-related) attack since these two-way information and control capabilities are built around microprocessor devices whose basic functions are controlled by software programming. Safeguarding large transformers and key substations from a potential physical attack by terrorists is a concern. Protection from a major geomagnetic disturbance event caused by solar storms is also a key issue, as prolonged, widespread electricity outages could result.[17]

Recent Events

High winds, especially when combined with precipitation from seasonal storms, can cause damage to electricity utility systems, resulting in service interruptions to large numbers of electricity customers. Data from various studies lead to cost estimates from storm-related outages to the U.S. economy at between $20 billion and $55 billion annually. Data also suggest the trend of outages from weather-related events is increasing.[18] The recent damage sustained to the electrical grid by Hurricane Sandy in New York and New Jersey, and difficulty in restoring electricity service underscore the age and fragility of the power system, and how electricity service might benefit from hardening and modernization of various power systems.

Selected Congressional Action from 112th Congress

Cyber Security and American Cyber Competitiveness Act of 2011 (S. 21, Reid)

The bill called for the enactment of bipartisan legislation to secure the United States against cyber attack, enhance American competitiveness, create jobs in the information technology industry, and protect the identities and sensitive information of American citizens and businesses by (1) enhancing the security and resiliency of U.S. government communications and information networks against cyber attack; (2) incentivizing the private sector to quantify, assess, and mitigate cyber risks to networks; (3) promoting investments in the American information technology sector; (4) improving the capability of the government and the private sector to assess cyber risks and prevent, detect, and respond to cyber attacks; (5) preventing and mitigating identity theft; (6) enhancing U.S. diplomatic capacity and international cooperation to respond to emerging cyber threats; (7) protecting and increasing the resiliency of U.S critical infrastructure and assets against cyber attacks. Introduced January 25, 2011; referred to the Senate Committee on Homeland Security and Governmental Affairs.

Secure High-Voltage Infrastructure for Electricity from Lethal Damage Act (H.R. 668, Franks)

Amends the Federal Power Act to authorize FERC, with or without notice, hearing, or report, to order emergency measures to protect the reliability of either the bulk-power system or the defense critical electric infrastructure[19] whenever the President issues a written directive or determination identifying an imminent grid security threat. Directs FERC to consult with governmental authorities in Canada and Mexico regarding implementation of emergency measures. Prescribes (1) implementation procedures; and (2) related cost recovery measures affecting owners, operators, or users of either the bulk-power system or the defense critical electric infrastructure. Directs FERC to require any owner, user, or operator of the domestic bulk-power system to implement measures to protect the system against specified vulnerabilities. Introduced February 11, 2011; referred to the House Energy and Commerce Subcommittee on Energy and Power.

Additional References

Failure to Act: The Economic Impact of Current Investment Trends in Electricity Infrastructure, American Society of Civil Engineers, http://www.asce.org/Infrastructure/ Failure-to-Act/Electricity/.

Introduction to NISTIR 7628 Guidelines for Smart Grid Cyber Security, The Smart GridInteroperability Panel, September 2010, http://www.nist.gov/ smartgrid/upload/ nistir7628_total.pdf.

The Future of the Electric Grid, Massachusetts Institute of Technology, December 5, 2011, http://mitei.mit.edu/system/files/Electric_Grid_Full_ Report.pdf.

Environmental Issues

Coal has long been the major fossil fuel used to produce electricity. However, burning coal results in environmental consequences, such as nitrogen oxides and sulfur emissions. Today, the burning of coal and other fossil fuels is also largely believed to be contributing to global climate change and its potentially damaging effects. Over the last 40 years, Congress has directed the Environmental Protection Agency (EPA) to reduce the potential health and environmental impacts of fossil fuel use by limiting emissions or other consequences of combustion processes.[20] These environmental regulatory requirements have been evolving in the last decade due to various challenges to EPA's implementation of federal laws.

Recent Events

New regulations under development at EPA would impose new requirements on coal-fired power plants. Some of these rules would be implemented at the federal level, while others would be implemented at the state level. They include the Cross-State Air Pollution Rule[21] (which replaced the Clean Air Interstate Rule); the Mercury and Air Toxics Standards (MATS) (also known as the Utility MACT) rule to reduce emissions of mercury, other metallic toxics, acid gases, and organic air toxics; the proposals to regulate coal combustion residues; and the Clean Water Act section 316(b) cooling water intake rule. However, only the Utility MACT rule is currently in effect.[22] EPA also issued standards for greenhouse gas emissions which would require all new power plants to restrict carbon dioxide emissions. EPA has yet to propose rules for GHG emissions from existing power plants, as is required by court order. Much attention has focused recently on the resulting finalization of these regulations, and their potential to contribute to the retirement of mostly small, older coal-burning power plants without modern environmental controls. Due to a general perception in the electric power industry that these new and pending environmental regulations present conflating requirements with unrealistic timeframes for compliance, the regulations have come to be referred to by the industry as the "train wreck" scenario due to a perception that a negative impact on reliability could result. Although environmental groups, and some in the electric power industry—mainly those with significant investments in nuclear or natural gas-fired generation—consider the concerns overstated.[23]

Selected Congressional Action from 112th Congress

Ensuring Affordable Energy Act (H.R. 153, Poe)

Prohibits any funds appropriated or otherwise available for the Administrator of the Environmental Protection Agency from being used to implement or enforce (1) a cap-and-trade program; or (2) any statutory or regulatory requirement pertaining to emissions of one or more greenhouse gases from stationary sources that is issued or becomes applicable or effective after January 1, 2011. Defines (1) "cap-and-trade program" as any regulatory program established after the date of enactment of the bill that provides for the sale, auction, or other distribution of a limited amount of allowances that permit the emission of one or more greenhouse gases; and (2) "greenhouse gas" to include carbon dioxide, methane, nitrous oxide, sulfur hexafluoride, hydrofluorocarbons, perfluorocarbons, or any other designated anthropogenic gas. Introduced February 1, 2011; referred to the Subcommittee on Energy and Power.

Stop the War on Coal Act of 2012 (H.R. 3409, Johnson)

Prohibits the Secretary of the Interior, before December 31, 2013, from issuing or approving any proposed or final regulation under the Surface Mining Control and Reclamation Act of 1977 that would (1) adversely impact employment in coal mines in the United States; (2) cause a reduction in revenue received by the federal government or any state, tribal, or local government, by reducing through regulation the quantity of coal in the United States that is available for mining; (3) reduce the quantity of coal available for domestic consumption or for export; (4) designate any area as unsuitable for surface coal mining and reclamation operations; or (5) expose the United States to liability for taking the value of privately owned coal through regulation.

Among many other actions, the bill amends the Clean Air Act to prohibit the Administrator of the Environmental Protection Agency from promulgating any regulation concerning, taking action relating to, or taking into consideration, the emission of a greenhouse gas (GHG) to address climate change. Excludes GHGs from the definition of "air pollutant" for purposes of addressing climate change. Repeals and nullifies a number of federal rules and EPA regulatory actions. Introduced November 14, 2011; referred to the Senate and referred to the Committee on Environment and Public Works.

Additional References

EPA Regulations and Electricity—Better Monitoring by Agencies Could Strengthen Efforts toAddress Potential Challenges, United States Government Accountability Office, July 2012, http://www.gao.gov/ assets/600/592542.pdf.

2012 State of Reliability, North American Electric Reliability Corporation, May 2012, http://www.nerc.com/files/2012_SOR.pdf.

Preparing for EPA Regulations: Working to Ensure Reliable and Affordable Environmental Compliance, The Regulatory Assistance Project, July 2012, http://www.raponline.org/document/ download/id/919.

End Notes

[1] See definition of "Balancing Authority" at http://www.eia.gov/tools/glossary/index.cfm?id=B.

[2] The U.S. electric industry comprises over 3,000 public, private, and cooperative utilities, and more than 1,000 independent power generators. See *Electricity Regulation In the US: A Guide*, Regulatory Assistance Project, March 2011, http://www.raponline.org/document/ download/id/645.

[3] The bulk power system makes it possible for utilities to engage in wholesale (sales for resale) electric power trade. U.S. Energy Information Administration, Overview—Power Transactions & Interconnected Networks, 2011, http://www.eia.gov/cneaf/electricity/page/ prim2/chapter7.html.

[4] "Services that ensure reliability and support the transmission of electricity from generation sites to customer loads. Such services may include load regulation, spinning reserve, non-spinning reserve, replacement reserve, and voltagesupport." See http://www.eia.gov/tools/ glossary/index.cfm.

[5] Includes electric utilities and independent power producers.

[6] U.S. Energy Information Administration, *Emissions of Greenhouse Gases in the United States 2008*, DOE/EIA-0573(2008), December 2009, http://www.eia.doe.gov/oiaf/1605/ggrpt/pdf/ 0573(2008).pdf.

[7] U.S. Energy Information Administration, *Electricity Net Generation: Total (All Sectors), 1949-2009*, Annual Energy Review 2009, Report No. DOE/EIA-0384(2009), July 19, 2010, http://www.eia.gov/emeu/aer/ elect.html.

[8] A commodity is an economic good, or a product available for shipment as a mass-produced, unspecialized product. See http://www.merriam-webster.com/dictionary/commodity.

[9] Power is the rate of producing, transferring, or using electricity. Power is measured in Watts and often expressed in kiloWatts (kW) or MegaWatts (MW).

[10] Electrical energy is the ability of an electric current to produce work, heat, light, or other forms of energy. It is measured in kiloWatt-hours.(kWh). Example: A 100kW generator operating for 1 hour produces 1 kWh of electrical energy. See http://www.eia.gov/tools/glossary/ index.cfm?id=E.

[11] California suffered through a series of electricity shortages caused mainly by market manipulation to decrease energy supplies and drive up electricity prices. See Staff Report, *Price Manipulation in Western Markets*, Federal Energy Regulatory Commission, Docket No. PA02-2-000, March 26, 2003, http://www.ferc.gov/legal/maj-ord-reg/land-docs/ Gelinas_at_a_glance2.pdf.

[12] See 119 FERC ¶ 63,013 (2007) at http://www.ferc.gov/industries/electric/indus-act/wec/ gaming-initial-decision.pdf.

[13] Office of Enforcement, *2012 Report on Enforcement*, Federal Energy Regulatory Commission, Docket No. AD0713-005, November 15, 2012, http://www.ferc.gov/legal/staff-reports/11-15-12-enforcement.pdf #xml=http://search.atomz.com/search/pdfhelper.tk?sp_o= 2,100000,0.

[14] EIA, *Fuel used in electricity generation is projected to shift over the next 25 years*, July 30, 2012, http://www.eia.gov/todayinenergy/detail.cfm?id=7310.

[15] "In the Reference case, the natural gas share of electric power generation increases from 24 percent in 2010 to 28 percent in 2035, while the renewables share grows from 10 percent to 15 percent. In contrast, the share of generation from coal-fired power plants declines. The historical reliance on coal-fired power plants in the U.S. electric power sector has begun to wane in recent years. Over the next 25 years, the share of electricity generation from coal falls to 38 percent, well below the 48-percent share seen as recently as 2008, due to slow growth in electricity demand, increased competition from natural gas and renewable generation, and the need to comply with new environmental regulations." See http://www.eia.gov/forecasts/aeo/pdf/0383(2012).pdf.

[16] CRS Report R41886, *The Smart Grid and Cybersecurity—Regulatory Policy and Issues*, by Richard J. Campbell

[17] Radiation or charged particles ejected into space weather from solar flare eruptions could cause geomagnetic storms. Periodically, these eruptions from the sun are powerful enough to disrupt the operations of electric power systems. Solar flares occur in cycles of almost 11 years. The last major solar flare eruption in 1989 caused blackouts across the Canadian province of Quebec for nine hours. But even greater storms can occur perhaps every 100 or more years. An event observed in 1921 demonstrates that although these events are rare, they are likely to occur again. The reoccurrence today of an event like the 1921 storm could potentially result in large-scale blackouts affecting more than 130 million people and would expose more than 350 transformers to the risk of permanent damage. It could take weeks or months to replace most of the damaged transformers.

Some of the larger units could take years to replace since there are no U.S. power transformer manufacturers, and a multi-year backlog exists for the larger units.

See Committee on the Societal and Economic Impacts of Severe Space Weather Events, National Research Council, *Severe Space Weather Events—Understanding Societal and Economic Impacts: A Workshop Report*, NationalAcademies, May 22-23, 2008, http://books.nap.edu/catalog.php?record_id=12507.

[18] CRS Report R42696, *Weather-Related Power Outages and Electric System Resiliency*, by Richard J. Campbell

[19] The bill defines the term 'defense critical electric infrastructure' as any infrastructure located in the United States used for the generation, transmission, or distribution of electric energy that—(A) is not part of the bulk-power system; and (B) serves a facility designated by the President (as both critical to the defense of the United States and vulnerable to a disruption of the supply of electric energy provided to such facility by an external provider), but is not owned or operated by the owner or operator of such facility.

[20] For example, with the Clean Air Act of 1970 (P.L. 91-604) and subsequent revisions, the Clean Water Act of 1972 (P.L.92-500), and the Resource Conservation and Recovery Act of 1976 (P.L. 94-580).

[21] On August 21, 2012, in a 2-1 decision, the D.C. Circuit Court of Appeals vacated and remanded the rule, finding that EPA's imposition of Federal Implementation Plans, without first giving the states an opportunity to develop their own plans, was unlawful. See CRS Report R41563, *Clean Air Issues in the 112th Congress*, by James E. McCarthy.

[22] The proposed regulations are discussed in the CRS Report R41914, *EPA's Regulation of Coal-Fired Power: Is a "Train Wreck" Coming?*, by James E. McCarthy and Claudia Copeland.

[23] CRS Report R42144, *EPA's Utility MACT: Will the Lights Go Out?*, by James E. McCarthy

In: The Transformation of Electrical Power: Key Issues
Editor: Dawson B. Munoz
ISBN: 978-1-62618-619-4

Chapter 2

PROSPECTS FOR COAL IN ELECTRIC POWER AND INDUSTRY*

Richard J. Campbell, Peter Folger and Phillip Brown

SUMMARY

For most of the twentieth century, the primary use of coal in the United States was for electric power generation, and for most of the history of power generation in the United States, coal has been the dominant fuel used to produce electricity. Even as recently as 2011, coal was the fuel used for almost 42% of power generation in the United States accounting for 93% of coal use. Industrial uses represented the remaining 7%. However, in April 2012, coal's share of the power generation market dropped to about 32% (according to Energy Information Administration statistics), equal to that of natural gas. Coal was the fuel of choice because of its availability and the relatively low cost of producing electricity in large, coal-burning power plants which took advantage of coal's low-priced, high energy content to employ economies of scale in steam-electric production. However, coal use for power generation seems to be on the decline, and the magnitude of coal's role for power generation is in question. Two major reasons are generally seen as being responsible: the expectation of a dramatic rise in natural gas supplies, and the impact of environmental regulations on an aging base of coal-fired power plants.

A recent drop in natural gas prices has been enabled by increasing supplies of natural gas largely due to horizontal drilling and hydraulic fracturing (i.e., fracking) of shale gas formations. If the production can be sustained in an environmentally acceptable manner, then a long-term, relatively inexpensive supply of natural gas could result. Decreased natural gas prices are lowering wholesale electricity prices, stimulating a major switch from coal to gas-burning facilities. The electric utility industry values diversity in fuel choice options since reliance on one fuel or technology can leave electricity producers vulnerable to price and supply volatility. However, an "inverse relationship" may be developing for coal vs. natural

---

* This is an edited, reformatted and augmented version of the Congressional Research Service Publication, CRS Report for Congress R42950, dated February 4, 2013.

gas as a power generation choice based on market economics alone, and policies which allow one fuel source to dominate may come at the detriment of the other.

Coal-fired power plants are among the largest sources of air pollution in the United States. More than half a dozen separate Clean Air Act programs could possibly be used to control emissions, which makes compliance strategy potentially complicated for utilities and difficult for regulators. Because the cost of the most stringent available controls, for the entire industry, could range into the tens of billions of dollars, some power companies have fought hard and rather successfully to limit or delay regulations affecting them, particularly with respect to plants constructed before the Clean Air Act Amendments of 1970 were passed. The expected retirement of approximately 27 GW of coal-fired capacity by 2016 has been reported to the Energy Information Administration (EIA) by coal plant owners and operators, accounting for approximately 8.5% of U.S. coal-fired capacity. While the costs of compliance with new Environmental Protection Agency regulations are a factor, several other issues are cited by coal plant owners and operators as contributing to these retirement decisions including the age of coal-fired power plants, flat to modest electricity demand growth, the availability of previously underutilized natural gas combined-cycle power plants, and the lower price of natural gas due to shale gas development. Even coal plants which have made significant modifications to meet existing EPA regulations are being closed or mothballed due to a combination of low natural gas prices, and the inability to sell power into other markets.

EIA expects coal to be a significant part of the U.S. power generation industry's future to well past 2030. But given price competition from natural gas, and emerging environmental regulations, that role will likely be smaller than in recent decades. Coal-fired generation is likely to face a challenging future.

INTRODUCTION

Coal has been the fuel most used for thermal energy since the early days of the Industrial Revolution. However, the mechanization of factories quickly brought a shift in coal use from heat to steam engines, and then to fuel steam turbines to power the electric motors of industry. The advent of electrical lighting and the electrification of buildings and residences drove even greater demand for electricity. For most of the twentieth century, the primary use of coal in the United States was for electric power generation.

In 2011, 93% of coal consumed in the United States was for electric power production, according to the Energy Information Administration (EIA). **Figure 1** shows the use of the four major types of coal produced in the United States, with bituminous and subbituminous coal dominating electric power generation.

The various types of coal differ mostly in carbon content and heating value:

- **Anthracite** has the highest carbon content (between 86% and 98%), and a heat value of about 15,000 British Thermal Units (BTUs) per pound.[1] Anthracite coal is a small part of the electric power market, and is mostly found in the Appalachian region of Pennsylvania.
- **Bituminous** is the most abundant form of coal in the United States, and is the type most commonly used to generate electricity. Bituminous coal has a carbon content of

approximately 45% to 86%, and a heat value between 10,500 BTUs and 15,500 BTUs.

- **Subbituminous** coal is mostly found in six western states and Alaska. It has a carbon content of between 35% and 45%, and a heat value of between 8,300 BTUs and 13,000 BTUs. Subbituminous coal generally has a lower sulfur content than other types of coal.
- **Lignite** has the lowest carbon content of the four types of coal generally used for electric power generation, averaging between 25% and 35%, and a high moisture and ash content. It also has the lowest heat value ranging between 4,000 BTUs and 8,300 BTUs.

ELECTRIC POWER GENERATION

For most of the history of power generation in the United States, coal has been the dominant fuel used to produce electricity. Even as recently as 2011, coal was used to fuel almost 42% of power generation in the United States, as shown in **Figure 2**. Coal has been the fuel of choice for many decades because of its wide availability, and the relatively low cost of producing electricity in large, coal-burning power plants. Coal's low-priced, high energy content enabled the building of power plants able to take advantage of economies of scale in steam-electric production. In a steam power plant, coal (or other fossil fuel) is burned to provide heat for turning water into steam in a boiler.[2] The steam is then forced under pressure into a steam turbine-driven generator which produces electricity. As of 2011, the U.S. coal-powered generation fleet consisted of 1,387 units with a nameplate capacity of almost 318 gigawatts[3] (GW)[4] representing approximately 30% of total U.S. generating capacity.[5]

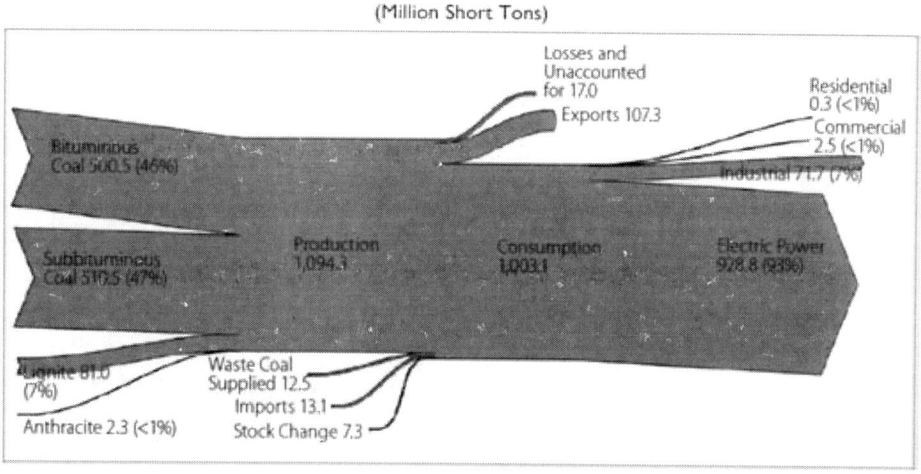

Source: U.S. Energy Information Administration. See http://www.eia.gov/totalenergy/data/annual/pdf/sec7_3.pdf.

Figure 1. Flow of U.S. Coal Consumption for 2011.

Net U.S. Generation 2011

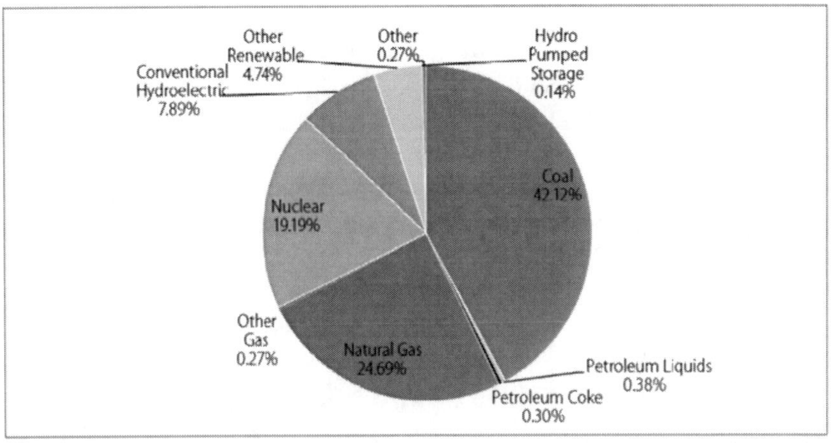

Source: CRS. Data from Energy Information Administration. Electric Power Monthly. Table 1.1, All sources of power.

Figure 2. Sources of Electric Power Generation.

Electric utilities and independent power producers[6] are the primary entities involved in U.S. electricity production. Together, these electric power sector participants have combined for an average annual consumption of almost 995 million tons of coal in the period from 2002 to 2011,[7] with electric utilities consuming between 60% to 70% of the total (primarily in base load[8] power plants).

The capital costs of power plant construction (i.e., the costs of finance and construction) vary according to the type (for example, coal or nuclear), and size of power plant. The cost for these plants to generate electricity is largely based on their cost of fuel, as well as operations and maintenance costs, other capital costs, and how the power plant is operated (for example, whether the facility is run as a base load or peaking[9] facility).[10] While the regulatory regime (i.e., under a rate-regulated or competitive environment), costs of transmission and distribution, and weather are strong determinants, the cost of fuel is often the major factor affecting electricity production especially for power plants which have already recovered their capital costs of construction.

Transportation costs for coal have historically been a major consideration in the choice of coal as a fuel. Barge transport is used where river access is available. But rail transportation is the most common mode of transit for coal as unit trains[11] offer the ability to transport large shipments to power plants on a regular basis. However, rail transportation can be costly.

> First, coal-burning electric power plants receive approximately 72 percent of their coal by rail. Second, rail transportation costs account for a sizable share of total delivered costs. While, on average, transportation costs account for approximately 20 percent of total [delivered] costs, they can reach as high as 59 percent on shipments of coal originating in the Powder River Basin.[12]

U.S. coal transportation costs are cited as a major reason why many utilities in the New England states have chosen to generate electricity with natural gas, or import coal from overseas as a lower cost alternative.[13]

Recent Changes in Markets for Electricity Fuels

The amounts of coal consumed for electricity generation from 2002 to 2011 are shown in **Figure 3**. While coal was the dominant fuel for electricity production in the period, the trend of increasing consumption seems to be slowing with a recent downturn in coal consumption for electricity beginning in about 2008. At the same time, natural gas use for electric power production is rising.

Power Plant Dispatch and Competitive Power Production

Many regions of the United States use competitive markets for power generation which sell into the wholesale market. These markets are generally administered by regional transmission organizations (RTOs) which have a goal of minimizing the cost of wholesale electricity to distribution utilities. The distribution utilities sell power to end-use (retail) consumers. In RTO markets, power plants bid against each other to serve the electricity demand of the wholesale market. Operators "dispatch" (or schedule the operation of) power plants based on the least costly choice, considering forecast loads and transmission system constraints for specific load increments at various times of the day or year. The most efficient power plants with the lowest heat rates (the efficiency with which a power plant turns fuel into electricity) are typically dispatched more, and lower efficiency, higher heat rate units are dispatched less.

Lower natural gas prices have caused wholesale electricity prices to decline, lowering potential revenues. However, many of the newer, more efficient natural gas combined cycle power generation plants have heat rates around 7,000 BTU/kWh (compared to today's coal plants with an overall average heat rate of 10,000 BTU/kWh). As natural gas prices have declined, natural gas plants have been dispatched more, and gross margins have increased with the spark spread (a measure of the potential profitability of a natural gas-fired power plant based on the difference between the price that a generator can obtain from selling one megawatt hour (MWh) of electricity and the cost of the natural gas needed to generate the MWh of electricity).

For coal plants, the lower cost of wholesale power has combined with the rising costs of coal to decrease their potential profitability. Older, less efficient coal power plants are especially at risk because wholesale prices of electricity may be inadequate for such facilities to operate economically.

Increasing supplies of natural gas are largely behind the switch in fuel for power generation. Natural gas prices have been volatile over the last few decades, ranging from less than $2 per million Btus (MMBTU) to over $18 per MMBTU.[14]

However, supplies of natural gas increased markedly beginning in 2008-2009 largely due to development of unconventional shale gas through improved drilling and simulation techniques. Horizontal drilling and hydraulic fracturing (i.e., fracking) have dramatically improved shale gas production, and the increased supplies of natural gas have caused prices to fall.

Figure 4 shows natural gas and coal fuel costs for electricity production from 2002 to 2012. The right-most part of the chart is based on monthly data, illustrating the recent drop in natural gas prices for electricity production while coal costs continue to inch higher. The drop in natural gas prices is a major factor in the recent switch from coal to gas-burning facilities, since the cost of fuel is a major determinant in the dispatch of electricity production. Switching from coal to gas generation has largely involved a switch from base load coal generation to intermediate load[15] natural gas combined cycle (NGCC) capacity. There are varying amounts of underutilized NGCC capacity available in various regions of the United States. That underutilized capacity resulted largely from the natural gas "bubble" of the 1980s and 1990s when natural gas prices were generally about $2 per MMBTU.

Fuel cost to generate electricity is the key in the decision to switch from coal to natural gas generation. If a power plant was 100% efficient, it would take 3.412 MMBTU to generate 1 MWh of electricity.

Assuming that an intermediate capacity NGCC plant is 44.78% efficient, it would consume 7.619 MMBTUs to generate 1 MWh of electricity. Similarly, a base load coal plant with an efficiency of 33.64% would use 10.412 MMBTUs to generate 1 MWh of electricity.[16] The cost of fuel to generate 1 MWh of electricity from coal has historically been lower than the cost to generate power from natural gas. The economics are now changing with coal prices rising and the drop in natural gas prices in 2011 down to levels where it now becomes more expensive to generate electricity from coal, especially when power plant efficiency is taken into account.

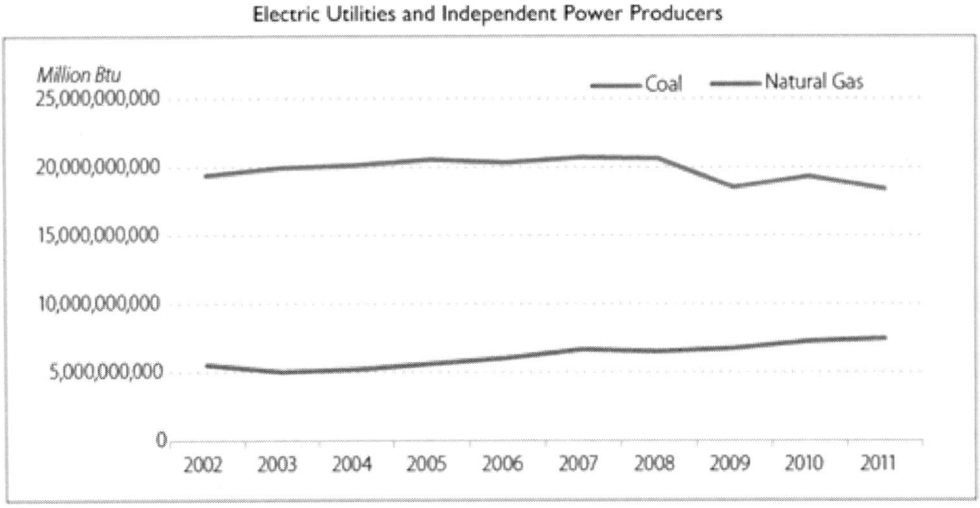

Source: CRS, data from U.S. Energy Information Administration (EIA), *Electric Power Monthly*, Table 2.1.A, "Coal: Consumption for Electricity Generation by Sector," at http://www.eia.gov/electricity/ monthly/pdf/epm.pdf, September 2012; CRS, data from U.S. Energy Information Administration (EIA), Electric Power Monthly, Table 2.4.A, "Natural Gas: Consumption for Electricity Generation by Sector," at http://www.eia.gov/electricity/ monthly/pdf/epm.pdf.

Figure 3. Fuel Consumption for Electricity Generation 2002-2011.

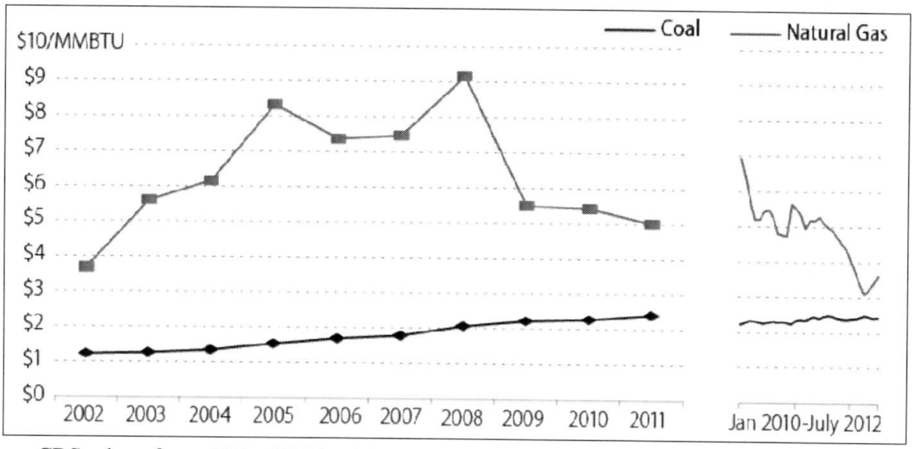

Source: CRS, data from EIA, "Table 4.2, Receipts, Average Cost, and Quality of Fuel, Electric Utilities" at http://www.eia.gov/electricity/monthly/pdf/epm.pdf. CRS. Data from EIA, Table 4.2, Receipts, Average Cost, and Quality of Fuel, Electric Utilities at http://www.eia.gov/electricity/monthly/pdf/epm.pdf.

Figure 4. Cost of Fuel for Electricity Generation, 2002-2011.

While the chart in **Figure 5** shows the expected rise and fall of power production with the seasons in the 2009 to 2012 timeframe, a look behind the numbers reveals the shift in generation as more power is being dispatched from higher efficiency natural gas-fired power plants.[17] During this time, coal's share of total U.S electricity generation has dropped and the use of natural gas to generate electricity has risen as newer, more efficient natural gas-fired generating capacity are being dispatched.[18] In April 2012, for the first time in history, the amount of electricity generation from natural gas equaled that of coal, according to EIA statistics, each with about 32% of the market.[19]

Many in the energy industry believe that a structural change in the economics of natural gas use has begun.

> While coal is projected to retain the largest share of the electricity generation mix through 2035, analyses included in the Annual Energy Outlook 2012 ... anticipate its share declining as more generation comes from natural gas and renewable technologies. Coal's role as the preeminent source of electricity generation in the United States has lessened in recent years, declining from 49% of total electricity generation in 2007 to 42% in 2011.... Projected fuel prices and economic growth are key factors influencing the future electricity generation mix. The price of natural gas, coal's chief competitor, has dropped significantly in recent years due to the increase in domestic production of natural gas.[20]

Nevertheless, there is still much debate as to whether the "shale gas revolution" will result in a predictable, long-term increase in supplies of natural gas at relatively low prices.

> Considerable uncertainty exists regarding the size of the economically recoverable U.S. shale gas resource base and the cost of producing those resources. Across four shale gas resource scenarios from the [EIA's] Annual Energy Outlook 2012, natural gas prices vary by about $4/MMBtu in 2035, demonstrating the significant impact that shale gas resource uncertainty has in determining future natural gas prices. This uncertainty exists

primarily because shale gas wells exhibit a wide variation in their initial production rate, rate of decline, and estimated ultimate recovery per well (or EUR, which is the expected cumulative production over the life of a well).[21]

If these previously unconventional shale gas resources can be economically developed and produced in an environmentally acceptable manner, then a sustained, relatively inexpensive supply of natural gas could persist.[22]

As long as natural gas prices are "relatively" high, development of resources will continue. When a surplus of natural gas supplies causes the price to drop, resource development activities solely focused on natural gas tend to curtail. But fracking of shale gas resources is changing the outlook for natural gas supplies, and the recovery of oil from wet shale plays is currently driving development. Even so, the expectation of lower, less volatile prices from increased supplies is encouraging the greater use of natural gas for electric power generation and other industrial uses. Natural gas prices are still subject to forces of supply and demand, with the cost of developing incremental supplies determining future supplies of the resource. Further development of the interstate pipeline system may also be needed if these new natural gas supplies are to reach markets.[23]

How the natural gas industry will accommodate the expected demands of the power generation industry for gas supplies is also uncertain. Power plants purchase coal under long-term contracts, amassing the fuel on-site in large, managed coal storage piles which may hold a 30-day supply. Most natural gas power generation plants are served directly by pipelines, and manage the fuel as more of a "just-in-time" resource, since power plants typically do not have large natural gas storage capabilities. The Federal Energy Regulatory Commission observes that based on recent supply issues, this new volume of expected demand for natural gas may require more longer term coordination and planning between the electricity and natural gas supply industries.[24]

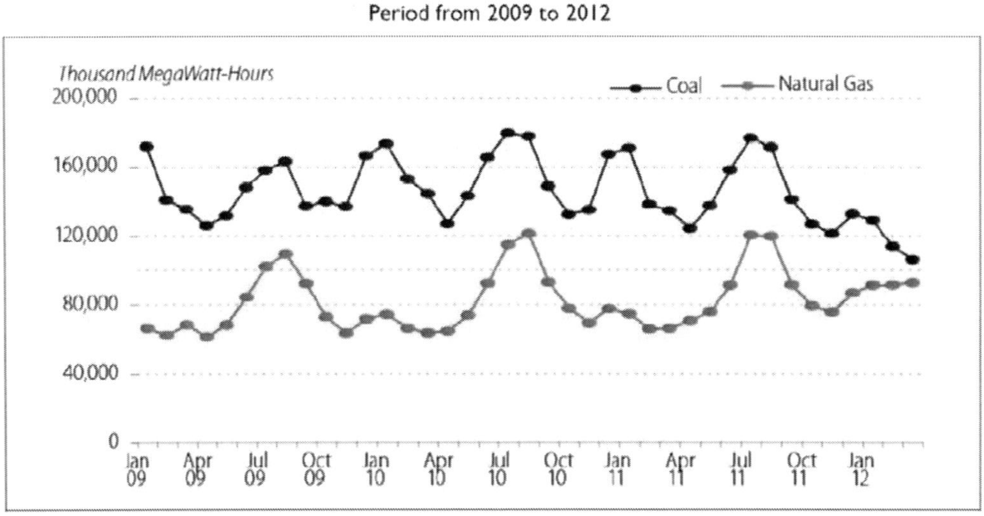

Source: CRS, data from EIA. See http://www.eia.gov/todayinenergy/ detail.cfm?id=6550.

Figure 5. Net Generation—Coal vs. Natural Gas.

Potential Coal Plant Retirements

All power plants are subject to retirement when they reach they reach the end of their useful service life. While older coal-fired power plants are usually well-maintained, they are generally not as efficient as newer power plants. As power plants age, they are generally upgraded to continue operations, but the least efficient plants may be retired. Other plants may be shifted from base load operations (in which they essentially operate around the clock) to less demanding intermediate or peaking schedules. The cost of building a power plant is generally recovered over the depreciable life of the asset, such that operations and maintenance (O&M) expenses become the major component of an older power plant's continuing costs. A major component of O&M is the cost of fuel, and the expectation of continued lower prices for natural gas is weighing on decisions concerning whether many older, less efficient coal power plants will be mothballed or closed altogether. However, the costs of modernizing older power plants to meet new regulatory requirements can be relatively high. When the cost of upgrades to meet new environmental requirements is considered along with (perhaps increasing) O&M expenses, many older coal power plants are likely to face outright retirement decisions.

Over the last 40 years, Congress has directed EPA to reduce the potential health and environmental impacts of fossil fuel use by limiting emissions by-products or other consequences of combustion processes.[25] These environmental regulatory requirements have been evolving in the last decade due to various challenges to the regulatory implementation of federal laws. Much attention has focused recently on the resulting finalization of some of these regulations, and their potential to contribute to the retirement of some coal-burning power plants.

In addition to being the largest source of electric power, coal-fired power plants are among the largest sources of air pollution in the United States. Under the Clean Air Act (CAA),[26] however, they have not necessarily been subject to stringent requirements: emissions and the required control equipment can vary depending on the location of the plant, when it was constructed, whether it has undergone major modifications, the specific type of fuel it burns, and, to some extent, the vagaries of EPA enforcement policies. More than half a dozen separate CAA programs could potentially be used to control emissions, which makes compliance strategy potentially complicated for utilities and difficult for regulators. Because the cost of the most stringent available controls, for the entire industry, could range into the tens of billions of dollars, power companies have fought hard and rather successfully to limit or delay regulations affecting them, particularly with respect to plants constructed before the Clean Air Act of 1970 was passed. However, EPA contends that such costs would be more than outweighed by societal benefits from reduced respiratory and other illnesses—"externalities" that are largely not currently accounted for in the price of electricity generation but which are estimated by EPA as costing the United States billions of dollars annually in health care and environmental costs.[27]

Some of the new rules under development at the EPA would be implemented at the federal level, while others would be implemented at the state level. They include the Cross-State Air Pollution Rule[28] (which replaced the Clean Air Interstate Rule); the Utility Maximum Achievable Control Technology (Utility MACT) rule to reduce emissions of mercury, other metallic toxics, acid gases, and organic air toxics; the wet ash classification rule for coal combustion residues; and the Clean Water Act Section 316(b) guidelines for once-through cooling water systems. However, only the Utility MACT rule is currently in effect. EPA also proposed standards for greenhouse gas (GHG) emissions which would require all new power plants to restrict carbon dioxide emissions.[29]

EPA has yet to propose rules for GHG emissions from existing power plants, as is required by court order. Due to a general perception in the electric power industry that these new and pending environmental regulations present conflating requirements with unrealistic timeframes for compliance, the regulations have come to be referred to by the industry as the "train wreck" scenario[30] due to a perception that a negative impact on reliability could result. However, environmental groups, and some in the electric power industry—mainly those with significant investments in nuclear or natural gas-fired generation—consider the concerns overstated.[31]

In 2011, the North American Reliability Corporation (which has oversight of the bulk electric grid for reliability purposes) analyzed the effects of the proposed EPA regulations for their potential to accelerate retirement of fossil-fired units.[32] The 2011 assessment was a revision of a previous analysis to take into account "updated assumptions determined from proposed rules, as well as other viewpoints on how the rule will ultimately be carried out (i.e., how states may implement regulations), to determine potential resource adequacy impacts." NERC noted that the expected retirements of electric generators are driven by many economic factors, "not simply the cost of pollution control equipment." However, NERC's assessment estimated that between 35 GW and 59 GW of generating capacity could be potentially retired as a result of the proposed EPA rules. NERC summarized its concerns as to the possible effects on reliability:

> Existing and proposed environmental regulations in the U.S. may significantly affect bulk power system reliability depending on the scope and timing of the rule implementation and the mechanisms in place to preserve reliability.[33]

For its part, the Department of Energy (DOE) estimated in 2011 that new air pollution, water, and waste disposal proposed rules could eventually result in the retirement of between 35 GW and 70 GW of coal-fired capacity.[34]

Electricity Reliability—State and Market Inputs

As retirement decisions are announced for growing amounts of coal capacity, questions are being asked about the cost and types of generation which may replace that capacity. EIA defines electric system reliability as "the degree to which the performance of the elements of the electrical system results in power being delivered to consumers within accepted standards and in the amount desired." Implicit in the definition is the necessity of generation adequacy to ensure that there is sufficient power plant capacity to meet projected demand, and additional reserves for exceptionally high demand periods. This includes voltage support (i.e., power injected into the grid to maintain voltage at acceptable limits) and ancillary services (i.e., services which support the reliable operation of the transmission system as it moves electricity from generating sources to retail customers) provided by retiring plants. Timely and appropriate replacement of retiring coal capacity would be essential to maintain reliability levels.

Cost recovery for a new power plant entails not only financial and performance risks, but also potential regulatory risk. In traditional cost of service states, adding a new power plant to a utility's rate base can raise customer rates.

While alternatives to potentially expensive new power plant construction may exist (such as demand response), if it appears that adding a new plant or plants to rate base will result in a significant rate increase, a reluctance may ensue on the part of state officials to authorize the addition especially in times of economic hardship. Similarly, investorowned utilities are wary of taking on too much debt for new power plant construction as this may decrease earnings and share price, possibly resulting in a downgrade of a company's financial rating.

In states with traditional cost of service ratemaking (in which electric utility rates are approved by state utility commissions), formal authorization is required before electric utilities can site, build, or place power plants into service thereby receiving recovery of costs in electricity rates. The need for new power plants is typically determined years in advance by a formal process estimating future demand and the ability of existing generation resources to meet that demand. A commonly used process is integrated resource planning (IRP), which considers a variety of supply and demand-side options, and assesses them against a common set of planning objectives and criteria, including environmental impact. The IRP objective is to help meet future customer demand by identifying the need for generating capacity, and determining the best mix of resources (including demand response) to fill the need at the lowest overall cost. Electric utilities often seek to shift the financial risk of building new power plants to ratepayers by obtaining regulatory approval to begin collecting the cost of the power plant in customer rates before the plant is in service (i.e., as construction work-in-progress).

Even in RTO markets where power generation is a competitive business, siting of new power plants is subject to state regulatory processes. However, in competitive markets, the financial risk of building a new power plant lies with the generator as they seek to bid to sell electricity into the RTO market. RTOs use markets to make operational decisions (such as generator dispatch) and to determine the provider(s) and prices for many of these services. RTOs do not own generation, but rely on price signals (based on real-time prices) to incentivize entry of new generation resources. Some RTOs employ forward capacity markets (up to three years in advance of expected needs) with auctions using either vertical demand curves (establishing quantity and allowing price to be set by the market) or sloped demand curves (regulating the price of new capacity) to contract for future capacity. Other RTOs use an administrative approach to determine capacity requirements and set prices. However, there is some debate as to how these market approaches work, and the resulting impact on prices. Other issues include whether forward capacity markets result in new capacity actually being built, whether the appropriate type of capacity (i.e., base load vs. intermediate load) is being incentivized and attracted, or even if the administratively-set or auction-determined capacity price is above the actual cost of new entry in these markets. The issue of prevailing high electricity prices in some areas (i.e., in load pockets) has led some states in RTO regions to consider authorizing the construction of power plants or bilateral contracting for electric power independent of RTO capacity market processes.

The expected retirement of approximately 27 GW of coal-fired capacity by 2016 has been reported to EIA by coal plant owners and operators, accounting for approximately 8.5% of U.S. coal-fired capacity. While the costs of compliance with new EPA environmental regulations are a factor, several other issues are cited by coal plant owners and operators as contributing to these retirement decisions including the age of coal-fired power plants, modest electricity demand growth, the availability of previously underutilized natural gas combined-cycle power plants,[35] and the lower price of natural gas due to shale gas development.[36] And some coal plants which have made significant modifications to meet existing EPA regulations are being closed or mothballed due to a combination of low natural gas prices, and either transmission congestion[37] or a lack of transmission options to sell power into other markets.[38]

The implications of coal-fired electric power plant retirements will likely play out on a regional basis, as EPA's rules go through inevitable challenges and modifications as the regulations are finalized.[39] Power generation in the United States varies according to local energy resources, and cost and availability of infrastructure for fuels transportation.[40] The Eastern (except New England) and Midwest regions of the United States are more coal-dependent than the Northeast or Western regions, as shown in **Figure 6**.

Several large regional transmission organizations[41] (RTOs) have undertaken studies of the impact of the new EPA rules. A major question being asked in these studies is whether there will be a risk to reliability based on the timing of requirements, and the scheduling and availability of equipment and personnel to accomplish retrofits of power plants to bring them into compliance with EPA regulations. The results of a study[42] by the Midwest ISO (MISO), the RTO serving 11 states and one Canadian province may provide insights into the implications of both EPA regulations and lower natural gas prices. There is approximately 70 GW of coal-fired capacity in the MISO market,[43] with an average coal plant age of 45 years.[44] The study identified potential retirements because of proposed EPA rules, a low natural gas price, and various levels of carbon mitigation costs. MISO summarized the results of its study noting that EPA regulations and lower natural gas prices would likely result in between 3 GW and 13 GW of coal capacity retirements, with MISO itself likely to see costs of between $32 billion and $33 billion due to EPA compliance retrofit costs, new capacity charges, system energy costs, and transmission system upgrades to maintain reliability.[45]

MISO does not own or build power plants. State utility commissions still control the process of new power plant construction in many states. States often require a utility to use integrated resource planning methods[46] to address system needs. If a determination is made that new generating capacity is needed to ensure reliability, then how to requisition that capacity becomes the responsibility of state public utility commissions or the RTO markets. The decision would likely be made according to how they view the effects of coal plant retirements on the levels and types (i.e., base load vs. intermediate or peak load power plants) of capacity that will be needed to serve electricity demands.[47] RTO markets generally rely on price signals[48] to provide an incentive for new generation to enter the market, or employ forward capacity markets with higher prices to encourage new generation to be built. Other "generation-like" resources (such as demand response,[49] energy efficiency, and conservation) would also likely be employed to reduce the need for building new capacity.

Given that significant underutilized NGCC exists in various U.S. regions, the possibility of further shifting from coal base load plants to natural gas intermediate capacity exists. A recent study[50] by the Massachusetts Institute of Technology in 2011 noted that the existing U.S. NGCC generation fleet had an average capacity factor[51] of approximately 41%, while its design capacity allowed such plants to operate at 85%. The MIT study looked at a scenario across selected regions of the United States which mimicked the "full dispatch" of existing natural gas combined cycle plants. The study concluded that under such a scenario (while noting that transmission constraints exist), there is "sufficient surplus NGCC capacity to displace roughly one-third of U.S. coal generation, reducing CO2 emissions from the power sector by 20%."[52]

Role of New and Alternative Power Generation Technologies

According to the EIA, coal is expected to be a key part of electricity generation in the United States well past the year 2030.[53] Electric utilities value diversity in power generation options since reliance on one fuel or technology can leave them vulnerable to price and supply volatility. However, the choices for new power plant technologies will likely be subject to proposed EPA rules limiting GHG emissions. The proposed rule published under the Clean Air Act's New Source Performance Standards program would limit emissions from new power plants to no more than 1,000 pounds of carbon dioxide for each megawatt-hour (MWh) of electricity produced. Today's NGCC generation emits on average about 800 pounds of carbon per MWh compared to current U.S. coal plants with average emissions of about 1,800 pounds of carbon dioxide per each MWh of electric power produced.[54] The proposed GHG rule would allow an exception for the approximately one dozen new coal plants that are permitted and beginning construction within one year, and would not apply to existing plants. Only new coal plants with carbon capture and storage (CCS) are likely to meet the proposed GHG regulations,[55] and CCS technology is estimated to add about 30% to a coal plant's cost. However, according to EIA, there are no new conventional coal plants expected to come on line after 2012; 2 GW of coal plant capacity with CCS is expected to enter service by about 2017.[56] New advanced NGCC plants entering the market in 2017 are expected to cost approximately $63 per MWh, while new advanced coal plants are estimated to cost $111 per MWh, and advanced coal with CCS is estimated at $139 per MWh.[57] Based on levelized cost estimates alone, new coal generation (even without CCS) may not be competitive with new NGCC given the built-in expectation of future fuel costs in the estimates.

The variability of most renewable electricity technologies means that these technologies offer limited opportunities for replacing coal generation, and these opportunities vary across the United States. Given that the best wind and solar resources are located far from population centers, opportunities to use wind power and solar power are likely dependent on development of transmission lines. Without major breakthroughs in technology efficiency or energy storage, growth of power generation from all renewable sources[59] is expected to be limited to no more than 16% of all generation by 2035.[60]

> Since load must be balanced on a continuous basis, units whose output can be varied to follow demand (dispatchable technologies) generally have more value to a system than less flexible units (non-dispatchable technologies) or those whose operation is tied to the availability of an intermittent resource.[61]

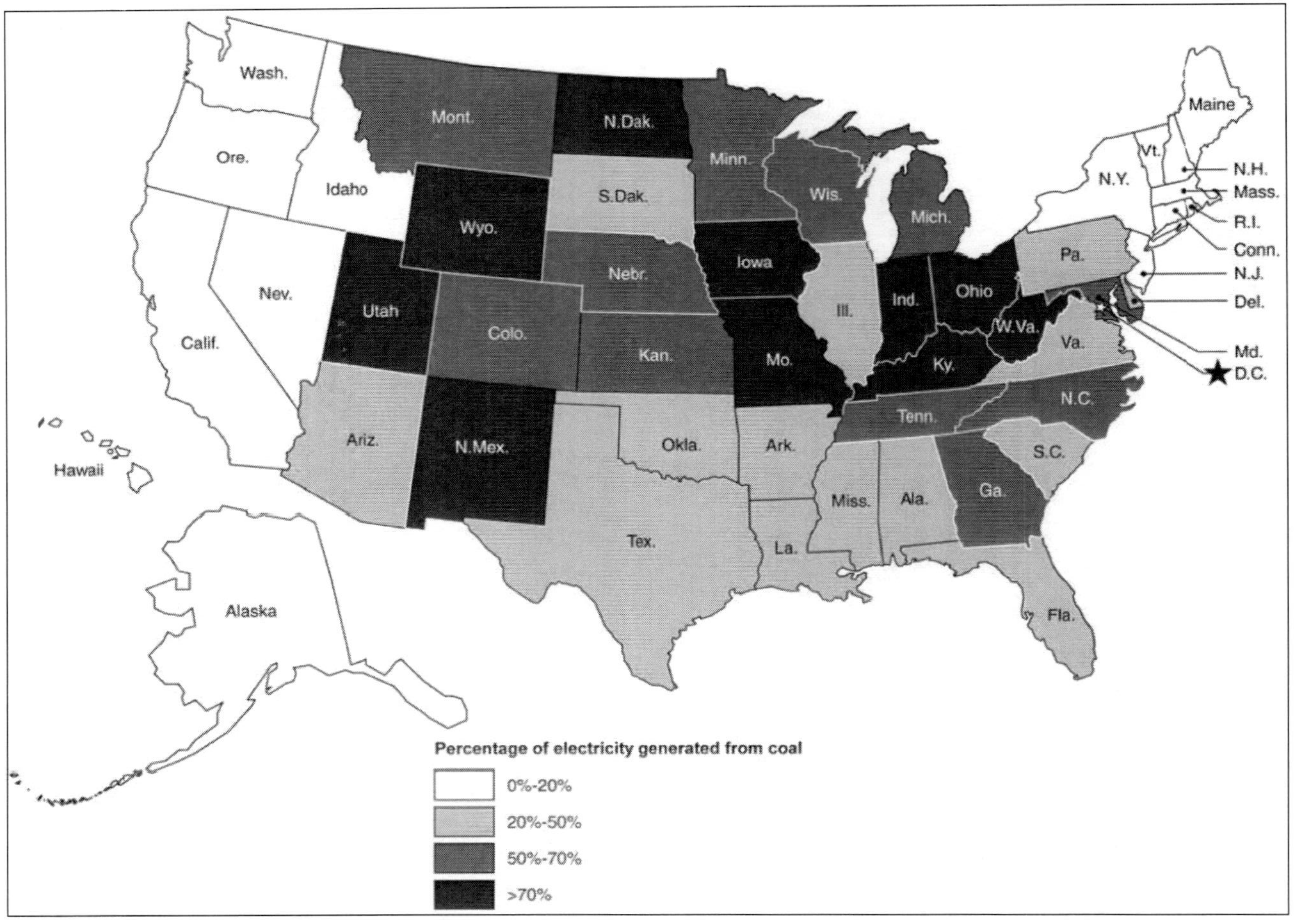

Source: Government Accountability Office. See http://www.gao.gov/assets/600/ 592542.pdf.

Figure 6. Percentage of Electricity Generated from Coal by State, 2010.

Development of joint renewable electricity and natural gas-fired power projects as "hybrid generation" is a possibility in certain regions. This could make variable renewable resources like wind and solar power more dispatchable, and would likely be dependent on state renewable portfolio standard[62] requirements.

Carbon Capture and Sequestration (CCS)58

Carbon capture and sequestration (or storage)—known as CCS—has attracted congressional interest as a measure for mitigating global climate change. Large amounts of carbon dioxide (CO_2) emitted from fossil fuel use in the United States are potentially available to be captured and stored underground using CCS, and prevented from reaching the atmosphere. Large, industrial sources of CO_2, such as coal-fired electricity-generating plants, are likely initial candidates for CCS because they are predominantly stationary, single-point sources. Electricity generation contributes over 40% of U.S. CO_2 emissions from fossil fuels. Currently, U.S. power plants do not capture large volumes of CO_2 for CCS.

To date, there are no commercial ventures in the United States that capture, transport, and inject industrial-scale quantities of CO_2 solely for the purposes of carbon sequestration. In 2012, the U.S. Department of Energy's CCS research, development, and deployment program embarked on commercial-scale demonstration projects for CO_2 capture, injection, and storage. The success or failure of these projects will likely bear heavily on the future outlook for widespread deployment of CCS technologies as a strategy for preventing large quantities of CO_2 from reaching the atmosphere while U.S. power plants continue to burn fossil fuels, mainly coal.

Nuclear power may be able to take over more of the base load generation role from coal, but the path forward for new nuclear generating units in such a role is uncertain. According to the Nuclear Energy Institute, as of 2012, five new nuclear power units are under construction, and maintaining nuclear energy's current share of 20% of total generation would require "building one reactor every year starting in 2016, or 20 to 35 new units by 2035."[63] The cost of building new nuclear power units may present a barrier to this goal,[64] as might concerns raised by the nuclear incident at Fukushima.

Given current technological options for power generation in the United States, the long-term expectation of lower natural gas prices in combination with existing and proposed EPA rules suggests that the option of coal use as a fuel for electric power generation faces an uncertain future.

INDUSTRIAL COAL CONSUMPTION[65]

In 2011 approximately 72 Million Short Tons (MST) of coal were used by the U.S industrial sector, which represented roughly 7% of total coal consumption in the country.[66] The industrial sector's share of U.S. coal demand has declined substantially since 1949, when the industrial sector accounted for nearly 45% of annual coal consumption (see **Figure 7**).[67] The decline of the industrial sector's share of coal consumption is generally a result of three developments since 1949: (1) reduced coal use for coke production and steel making, (2) reduced coal use for industrial process applications, and (3) increased coal use for electric power production, which now represents more than 90% of U.S. coal demand.[68]

Coal has a number of industrial uses including steel making, cement production, combined heat and power (CHP), and process heating. Additionally, coal is used by a number of industrial operations including alumina refineries, chemical producers, pharmaceutical companies, and paper manufacturers.[69] According to the World Coal Institute, there are thousands of products that include coal or coal by-products: soap, aspirins, solvents, dyes, plastics, fibers, and specialty products (e.g., activated carbon, carbon fiber, and silicon metal).[70]

The Energy Information Administration classifies industrial sector coal consumption into three separate categories: (1) Coke Plants, (2) Combined Heat and Power (CHP), and (3) Non-CHP. **Table 1** indicates the amount of coal consumed in 2011 for each of these categories.

Issues for Congress

Given the expected increase in natural gas supplies and expected shift in electricity production from coal to natural gas as a primary fuel, the issue of what level of coal generation should be preserved for power generation in the United States may be a question Congress will be faced with. While U.S. coal currently used for power generation may have options for markets overseas, the domestic future of coal power generation is clouded by both market prices and environmental regulations. The United States is believed to have a bountiful supply of coal. But how much or whether the future of coal will include power generation may be a question of national energy policy.

Congress may be also confronted with the question of whether an "all-of-the above" strategy for energy is viable. The electric utility industry values diversity in fuel choice options since reliance on one fuel or technology can leave electricity producers vulnerable to price and supply volatility. However, an "inverse relationship" is developing for coal vs. natural gas as a power generation choice based on market economics alone, and policies which allow one fuel source to dominate may come to the detriment of the other.

Table 1. 2011 Industrial Sector Coal Consumption
(Million Short Tons)

Coke Plants	CHP	Non-CHP	Total
21.4	24.7	25.6	**71.7**

Source: Energy Information Administration.

Congress may also face a decision as to whether coal-fired power generation can be a viable technology option for the future. New advanced NGCC plants entering the market in 2017 are expected to cost approximately $63 per MWh, while new advanced coal plants are estimated to cost $111 per MWh, and advanced coal with CCS is estimated at $139 per MWh. The cost difference between these technologies, and proposed GHG emissions limits, raise concerns over coal's future as a viable fuel choice for power generation. It is not clear (at this point in time) whether research and development will result in new technologies able to reduce the differential.

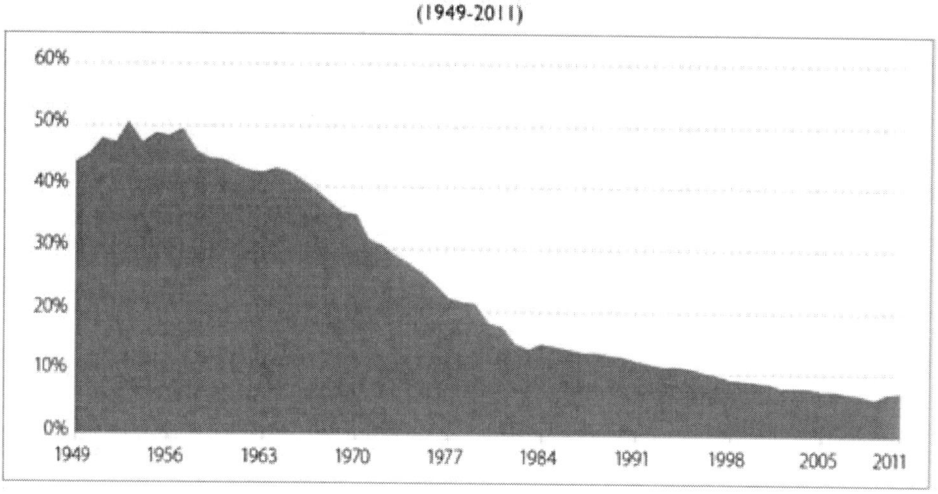

Source: Energy Information Administration.

Figure 7. Industrial Sector Share of Total U.S. Coal Consumption.

End Notes

[1] The heating value of any fuel is the energy released per unit mass or per unit volume of the fuel when the fuel is completely burned. Higher heat value fuels liberate more energy per unit of mass or volume.

[2] A simple cycle natural gas power plant burns natural gas in a combustion turbine (like a jet engine) to turn a generator which produces electricity. When used in a combined cycle mode, the hot exhaust gases from the combustion turbine are used to generate steam to turn a steam turbine to efficiently create additional electricity.

[3] One billion watts or one thousand megawatts.

[4] See http://www.eia.gov/todayinenergy/detail.cfm?id=7290.

[5] See http://www.eia.gov/totalenergy/data/annual/showtext.cfm?t=ptb0811a.

[6] A corporation, person, agency, authority, or other legal entity or instrumentality that owns or operates facilities for the generation of electricity for use primarily by the public, and that is not an electric utility. See http://www.eia.gov/tools/glossary/index.cfm?id=I#ind_pwr_prod.

[7] EIA, *Table 2.1.A. Coal: Consumption for Electricity Generation*, 2002-2011, http:// www. eia.gov/electricity/ monthly/pdf/epm.pdf.

[8] A plant, usually housing high-efficiency steam-electric units, which is normally operated to take all or part of the minimum load of a system, and which consequently produces electricity at an essentially constant rate and runs continuously. These units are operated to maximize system mechanical and thermal efficiency and minimize system operating costs. EIA, *EIA Glossary "Base Load Plant"*, http://www.eia.gov/tools/glossary/ index.cfm.

[9] A plant usually housing old, low-efficiency steam units, gas turbines, diesels, or pumped-storage hydroelectric equipment normally used during the peak-load periods. See http://www.eia.gov/tools/glossary/index.cfm.

[10] Black and Veatch, *Cost Report—Cost and Performance Data for Power Generation Technologies*, National Renewable Energy Laboratory, February 2012, http://bv.com/docs/ reports-studies/nrel-cost-report.pdf.

[11] Typically, all the freight cars on a "unit train" are filled with coal being shipped directly to a single destination.

[12] See http://www.eia.gov/coal/transportationrates/.

[13] See ISO New England, "Scenario Analysis Project—Long Term Forecast of Oil, Natural Gas and Coal Prices in New England," 2007, http://www.isone.com/committees/comm_wkgrps/ othr/sas/mtrls/apr22007/fuel-price-forecast.pdf.

[14] See Energy Information Administration, Office of Oil and Gas, *An Analysis of Price Volatility in Natural Gas Markets*, "Figure 1: Daily Henry Hub Spot Price, 1994-2006," August 2007, http://www.eia.gov/pub/oil_gas/natural_gas/feature_articles/2007/ngprivolatility/ngprivolatility.pdf.

[15] The range from base load to a point between base load and peak load.

[16] Sandy Fielden, "The Economics of Coal-to-Gas Switching," *Power Magazine*, August 21, 2012, http://www.powermag.com/business/The-Economics-of-Coal-to-Gas-Switching_ 4880_p4.html.

[17] EIA, "Natural Gas Spot Prices near 10-Year Lows Amid Warm Weather and Robust Supplies," February 1, 2012, http://www.eia.gov/todayinenergy/detail.cfm?id=4810.

[18] "Coal generation decreased 29 billion kilowatt-hours from March 2011 to March 2012, while natural gas generation increased 27 billion kilowatt-hours during the same time period. In March 2012, coal's share of total generation was34% compared to natural gas at 30%." See EIA, "U.S. Coal's Share of Total Net Generation Continues to Decline," June 5, 2012, http://www.eia.gov/todayinenergy/detail.cfm?id=6550.

[19] U.S. Energy Information Administration, *Net Generation by Energy Source: Total (All Sectors), 2002-June 2012*, July 2012, http://www.eia.gov/electricity/monthly/epm_table_ grapher.cfm?t=epmt_1_1.

[20] EIA, "Fuel Used in Electricity Generation Is Projected to Shift over the Next 25 Years," July 30, 2012, http://www.eia.gov/todayinenergy/detail.cfm?id=7310.

[21] EIA, Projected Natural Gas Prices Depend on Shale Gas Resource Economics," August 27, 2012, http://www.eia.gov/todayinenergy/detail.cfm?id=7710.

[22] For further discussion of natural gas production, see CRS Report R42814, *Natural Gas in the U.S. Economy: Opportunities for Growth*, by Robert Pirog and Michael Ratner

[23] Jack Buehrer, *Natural Gas Boom Drives Pipeline Upgrades*, Engineering News Record, May 7, 2012, http://enr.construction.com/infrastructure/power_industrial/2012/0507-shale-boomhttpprimisphmsadotgov constructionconstphmsahtm.asp.

[24] Federal Energy Regulatory Commission, *Natural Gas - Electric Coordination*, August 2012, http://www.ferc.gov/industries/electric/indus-act/electric-coord.asp.

[25] For example, with the Clean Air Act Amendments of 1970 (P.L. 91-604) and subsequent revisions, the Clean Water Act of 1972 (P.L.92-500), and the Resource Conservation and Recovery Act of 1976 (P.L. 94-580).

[26] The Clean Air Act, codified as 42 U.S.C. 7401 et seq., seeks to protect human health and the environment from emissions that pollute ambient, or outdoor, air. It requires the Environmental Protection Agency to establish minimum national standards for air quality, and assigns primary responsibility to the states to assure compliance with the standards. Areas not meeting the standards, referred to as "nonattainment areas," are required to implement specified air pollution control measures. The Act establishes federal standards for stationary and mobile sources of air pollution and their fuels and for sources of 187 hazardous air pollutants, and it establishes a cap-and-trade program for the emissions that cause acid rain. It establishes a comprehensive permit system for all major sources of air pollution. It also addresses the prevention of pollution in areas with clean air and protection of the stratospheric ozone layer. For more information, see CRS Report RL30853, *Clean Air Act: A Summary of the Act and Its Major Requirements*, by James E. McCarthy, Claudia Copeland, and Linda-Jo Schierow.

[27] USEPA Office of Air and Radiation, *The Benefits and Costs of the Clean Air Act: 1990 to 2020*, August 2010, http://www.epa.gov/air/sect812/aug10/fullreport.pdf.

[28] On August 21, 2012, in a 2-1 decision, the court vacated and remanded the rule, finding that EPA's imposition of Federal Implementation Plans, without first giving the states an opportunity to develop their own plans, was unlawful. See CRS Report R41563, *Clean Air Issues in the 112th Congress*, by James E. McCarthy. By vacating the rule, the Court determined that the Clean Air Interstate Rule would still apply. But that rule was itself vacated, making future implementation of any air transport rule unclear.

[29] Environmental Protection Agency, *Carbon Pollution Standard for New Power Plants*, http://epa.gov/carbonpollutionstandard/.

[30] These regulations are discussed in separate CRS Reports: CRS Report R41914, *EPA's Regulation of Coal-Fired Power: Is a "Train Wreck" Coming?*, by James E. McCarthy and Claudia Copeland, and CRS Report R42570, *Proposals to Amend RCRA: Analysis of Pending Legislation Applicable to the Management of Coal Combustion Residuals*, by Linda Luther.

[31] CRS Report R42144, *EPA's Utility MACT: Will the Lights Go Out?*, by James E. McCarthy.

[32] North American Electric Reliability Corporation, *Potential Impacts of Future Environmental Regulations—Extracted from the 2011 Long-Term Reliability Assessment*, November 2011, http://www.nerc.com/files/EPA%20Section.pdf.

[33] Ibid.

[34] Debra McCown, "EPA Regulations for Coal-Fired Power Plants Could Force Shut Downs," *Bristol Herald Courier*, May 24, 2011, http://www2.tricities.com/business/2011/may/25/ wood-gives-dire-warning-due-epa-regulations-coal-f-ar-1062322/.

[35] An electric generating technology in which electricity is produced from otherwise lost waste heat exiting from one or more gas (combustion) turbines. The exiting heat is routed to a conventional boiler or to a heat recovery

steam generator for utilization by a steam turbine in the production of electricity. This process increases the efficiency of the electric generating unit. See http://www.eia.gov/tools/glossary/index.cfm.

36 EIA, *27 GigaWatts of Coal-Fired Capacity to Retire over Next Five Years*, July 27, 2012, http://www.eia.gov/todayinenergy/detail.cfm?id=7290#.

37 A condition that occurs when insufficient transfer capacity is available to implement all of the preferred schedules for electricity transmission simultaneously. See http://www.eia.gov/ tools/glossary/index.cfm.

38 Gib Snyder, "An Inside Look at the Process for Mothballing," *The Observer*, September 9, 2012, http://www.observertoday.com/page/content.detail/id/575844/An-inside-look-at-the-process-for-mothballing.html?nav=5047.

39 InsideEPA, *EPA Urges Appellate Court To Rehear Ruling Vacating Cross-State Air Rule*, October 5, 2012, http://insideepa.com/201210052412215/EPA-Daily-News/Daily-News/ epa-urges-appellate-court-to-rehear-ruling-vacating-cross-state-air-rule/menu-id-95.html.

40 "In New England, most of the coal that will be burned ... will likely originate in South America or Indonesia, where the chemical composition of the coal is less harmful to the environment. Western U.S. coals also have a favorable chemical composition, but it is currently too expensive to deliver coal from the Rocky Mountains to New England relative to foreign supplies." ISO New England, *Scenario Analysis Project—Long Term Forecast of Oil, Natural Gas and Coal Prices In New England*, 2007, http://www.isone.com/committees/comm_wkgrps/othr/sas/mtrls/apr22007/fuel-price-forecast.pdf.

41 The purpose of a Regional Transmission Organization is to ensure the efficient and reliable delivery of power across large areas. RTOs facilitate and promote efficiency in wholesale electricity markets and ensure that electricity consumers pay the lowest price possible for reliable service by removing transmission barriers between buyers and sellers. See http://www.pjm.com/about-pjm/learning-center/pjm-overview/energy-basics/what-is-a-rto.aspx.

42 MISO, *EPA Impact Analysis- Impacts from the EPA Regulations on MISO*, October 2011, https://www.midwestiso.org/Library/Repository/Study/MISO%20EPA%20Impact%20Analysis.pdf. (MISO EPA)

43 Of the 70,000 MW of coal-fired capacity in the MISO market, less than half does not have plans for SO2 controls. Furthermore, 38% have no SO2 controls or NOx controls, and 38% have no SO2 controls or Fabric Filters. (MISO EPA)

44 The life expectancy of coal-fired power plants from a cost recovery basis is generally estimated to be 40 years.

45 MISO EPA.

46 "Integrated resource planning [IRP] is a crucial element for success in a constantly changing business and regulatory environment and is based on comprehensive, holistic and risk aware analysis. The integrated approach considers a broad spectrum of feasible supply and demand-side options and assesses them against a common set of planning objectives and criteria, including environmental impact. The IRP objective is to help meet future customer demand by identifying the need for generating capacity and determining the best mix of resources to fill the need. The capacity gap is the difference between the projected firm (or known) requirements and existing firm supply." See TVA, *TVA's Environmental and Energy Future*, Integrated Resource Plan, March 2011, p. 27, http://www.tva.gov/ environment/reports/irp/pdf/Final_IRP_Ch1.pdf.

47 Many electric utilities are required by states to undertake long-range planning. "Integrated resource planning helps electric utilities choose the best resource options to generate electricity and other options to meet customer expectations for energy services. Increasing competition, changing technologies, and environmental concerns are among the many issues utilities must consider when developing their plans." See "Purpose of and Need for Integrated Resource Planning" at http://www.tva.gov/environment/reports/ energyvision2020/ev2020_vol1ch01.pdf.

48 Shortages of capacity cause electricity prices to rise in a competitive market.

49 Demand response (also known as load response) is end-use customers reducing their use of electricity in response to power grid needs, economic signals from a competitive wholesale market or special retail rates. See http://www.pjm.com/markets-and-operations/demand-response.aspx.

50 Massachusetts Institute of Technology, *The Future of Natural Gas*, 2011, http://mitei.mit.edu/ system/files/NaturalGas_Report.pdf. (MIT Study)

51 The ratio of the electrical energy produced by a generating unit for the period of time considered to the electrical energy that could have been produced at continuous full power operation during the same period. See http://www.eia.gov/tools/glossary/index.cfm.

52 MIT Study, op. cit.

53 "In the Reference case, the natural gas share of electric power generation increases from 24 percent in 2010 to 28 percent in 2035, while the renewables share grows from 10 percent to 15 percent. In contrast, the share of generation from coal-fired power plants declines. The historical reliance on coal-fired power plants in the U.S. electric power sector has begun to wane in recent years. Over the next 25 years, the share of electricity generation from coal falls to 38 percent, well below the 48-percent share seen as recently as 2008, due to slow growth in electricity demand, increased competition from natural gas and renewable generation, and the need to comply with new environmental regulations." See http://www. eia.gov/forecasts/aeo/pdf/0383(2012).pdf.

54 40 C.F.R. § Part 60. See http://epa.gov/carbonpollutionstandard/pdfs/20120327proposal.pdf.

[55] "New coal-, coal refuse-, oil- and petroleum coke-fired boilers and IGCC [Integrated Gasification Combined Cycle] units should also be able to meet this standard by employing carbon capture and storage (CCS) technology." Ibid.

[56] Diane Kearney, *Coal Projections from the Annual Energy Outlook 2011 Early Release*, EIA, April 6, 2011, http://www.stb.dot.gov/stb/docs/RETAC/2011/April/EIA%20annual% 20energy%20outlook%202011.pdf.

[57] Total levelized system cost. "Levelized cost is often cited as a convenient summary measure of the overall competiveness of different generating technologies. It represents the per-kilowatthour cost (in real dollars) of building and operating a generating plant over an assumed financial life and duty cycle. Key inputs to calculating levelized costs include overnight capital costs, fuel costs, fixed and variable operations and maintenance (O&M) costs, financing costs, and an assumed utilization rate for each plant type. The importance of the factors varies among the technologies." See EIA, *Levelized Cost of New Generation Resources in the Annual Energy Outlook 2012*, Annual Energy Outlook 2012, July 12, 2012, http://www. eia.gov/forecasts/aeo/electricity_generation.cfm. (EIA NewGen)

[58] See CRS Report R42532, *Carbon Capture and Sequestration (CCS): A Primer*, by Peter Folger, and CRS Report R42496, *Carbon Capture and Sequestration: Research, Development, and Demonstration at the U.S. Department of Energy*, by Peter Folger.

[59] Includes conventional hydroelectric, geothermal, wood, wood waste, all municipal waste, landfill gas, other biomass, solar, and wind power.

[60] EIA, *Annual Energy Outlook 2012, Table A9*, DOE/EIA-0383(2012), May 2012, http://www. eia.gov/forecasts/ aeo/pdf/0383%282012%29.pdf.EIA, Table A9, Electric Generating Capacity. Annual Energy Outlook.

[61] EIA NewGen.

[62] Renewable Portfolio Standards are state programs requiring a percent of energy sales or installed capacity to come from renewable resources.

[63] Nuclear Energy Institute, *New Nuclear Energy Facilities*, 2012, http://www.nei.org/keyissues/ newnuclearplants/.

[64] The two units being built by the Southern Company at the Vogtle station in Georgia were originally estimated to cost $14 billion. See Ray Henry, "Building Costs Rise at U.S. Nuclear Sites," *South Coast Today.com*, July 15, 2012, http://www.southcoasttoday.com/apps/pbcs.dll/article?AID=/20120715/NEWS/207150345/-1/ NEWSMAP.

[65] This section was written by Phillip Brown.

[66] Energy Information Administration, *Annual Energy Review*, September 27, 2012.

[67] Ibid.

[68] Ibid.

[69] "The Coal Resource: A Comprehensive Overview of Coal," World Coal Institute, 2009.

[70] Ibid.

In: The Transformation of Electrical Power: Key Issues
Editor: Dawson B. Munoz
ISBN: 978-1-62618-619-4

Chapter 3

U.S. RENEWABLE ELECTRICITY GENERATION: RESOURCES AND CHALLENGES*

Phillip Brown and Gene Whitney

SUMMARY

The United States faces important decisions about future energy supply and use. A key question is how renewable energy resources might be used to meet U.S. energy needs in general, and to meet U.S. electricity needs specifically. Renewable energy sources are typically used for three general types of applications: electricity generation, biofuels/bioproducts, and heating/cooling. Each application uses different technologies to convert renewable energy sources into usable products. The literature on renewable energy resources, conversion technologies for different applications, and economics is massive. This report focuses on electricity generation from renewable energy sources. In 2010, renewable sources of energy were used to produce almost 11% (7% from hydropower and 4% from other renewables) of the 4 million gigawatthours of electricity generated in the United States.

This report provides a summary of U.S. electricity generation potential from wind, solar, geothermal, hydroelectric, ocean-hydrokinetic, and biomass sources of renewable energy. The focus of this report is twofold: (1) provide an assessment of U.S. renewable electricity generation potential and how renewables might satisfy electric power sector demand, and (2) discuss challenges, issues, and barriers that might limit renewable electricity generation deployment.

Data sources from 15 different organizations were reviewed to derive estimates of electricity generation potential. One key finding is that there exists no uniform national assessment of renewable electricity generation potential. No standard methods or set of assumptions are used to estimate renewable electricity generation potential. So even existing assessments for individual energy sources are difficult to compare objectively. In order to compare various estimates on an equivalent basis, CRS engaged experts in each renewable

* This is an edited, reformatted and augmented version of a Congressional Research Service publication, CRS Report for Congress R41954, from www.crs.gov, dated August 5, 2011.

energy resource area to help normalize electricity generation potential estimates into a common metric: gigawatthours per year.

After surveying, researching, and normalizing all of the third-party electricity generation estimates, results indicate that renewable energy sources may, in principle, have the potential to satisfy a large portion of U.S. electricity demand. However, a number of potential barriers to large-scale deployment exist, including cost, power system integration, intermittency and variability, land requirements, transmission access, possible limits to the availability of key materials and resources, certain environmental impacts, specialized infrastructure requirements, and policy issues. Ultimately, the amount of renewable electricity generation in the U.S. may be dependent on the ability to address these deployment barriers. The Energy Information Administration projects that U.S. renewable electricity generation will increase from 11% today to between 14% and 15% in 2035.

As Congress considers policy options associated with increasing renewable electricity generation, policy makers may assess potential benefits such as emissions reduction, job creation, and global competitiveness, along with possible risks and consequences such as electricity cost and price increases, electricity delivery reliability, and environmental impacts associated with large-scale deployment of renewable electricity generation technologies.

INTRODUCTION

The U.S. energy sector is large and complex. Multiple energy sources, including fossil, nuclear, and several renewable sources, are used to produce energy products for multiple demand sectors (transportation, electricity, industrial, and residential/commercial). Today, fossil fuels are the dominant sources of energy, comprising 83% of total U.S. primary energy supply. Renewable energy sources, which can be used to generate electricity, produce liquid transportation fuels, and provide heating and cooling for industrial and residential/commercial sectors, provided 8% of total U.S. primary energy supply in 2009 (see *Figure 1*).

The largest source of energy demand in the United States is the electric power sector, which consumed just over 40% of total U.S. energy supply in 2009. The U.S. electric power sector generates approximately 4 million gigawatthours of electricity each year. Like the total U.S. energy sector, electricity generation is dominated (89%) by fossil fuels and nuclear power.

Renewable electricity generation, including hydro, wind, solar, geothermal, and biomass, contributed 11% of total U.S. electric power in 2009 (*Figure 2*).[1] Most U.S. renewable generation comes from conventional hydropower, which has limited growth potential. Other renewable electricity sources constitute about 4% of U.S. generation, but have been growing more rapidly.

The purpose of this report is to analyze the prospects, opportunities, and challenges for renewable energy sources to increase their contribution to the electric power sector.

There is growing interest in increasing the amount of renewable electricity generation to reduce the amount of fossil fuel consumption for U.S. electric power. That interest is driven by concerns about greenhouse gas emissions, the realization that economically recoverable fossil fuel supplies are ultimately finite, and the desire to position the United States as a global leader for renewable energy technology and manufacturing.[2]

(Values are in Quadrillion Btu and Percentage of Total)

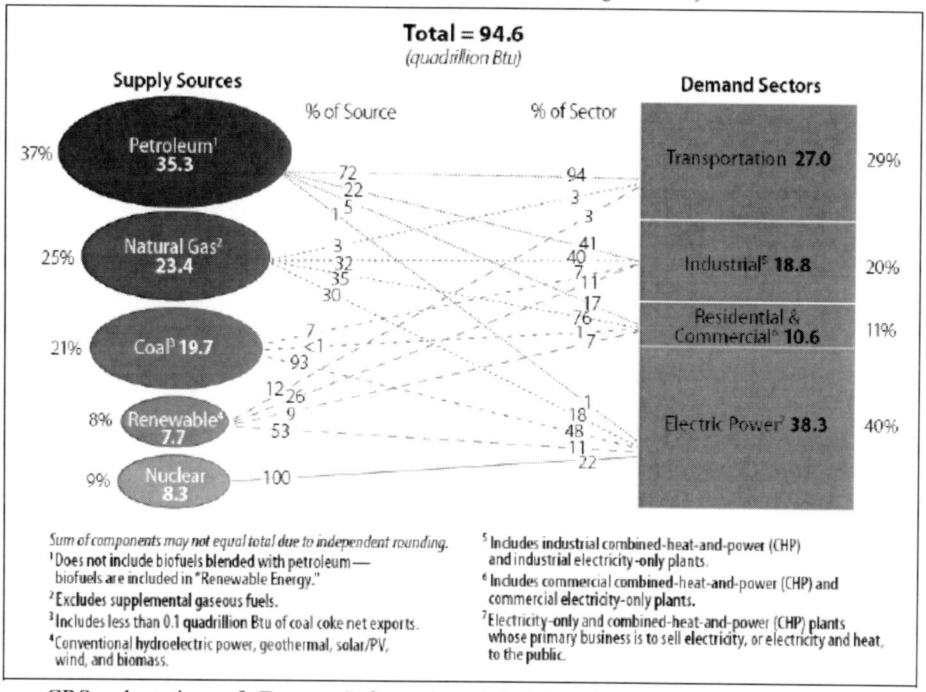

Source: CRS adaptation of Energy Information Administration, Annual Energy Review 2009, http://www.eia.doe.gov/totalenergy/data/annual/pdf/pecss_ diagram_2009.pdf.

Figure 1. U.S. Primary Energy Flow by Supply Source and Demand Sector, 2009.

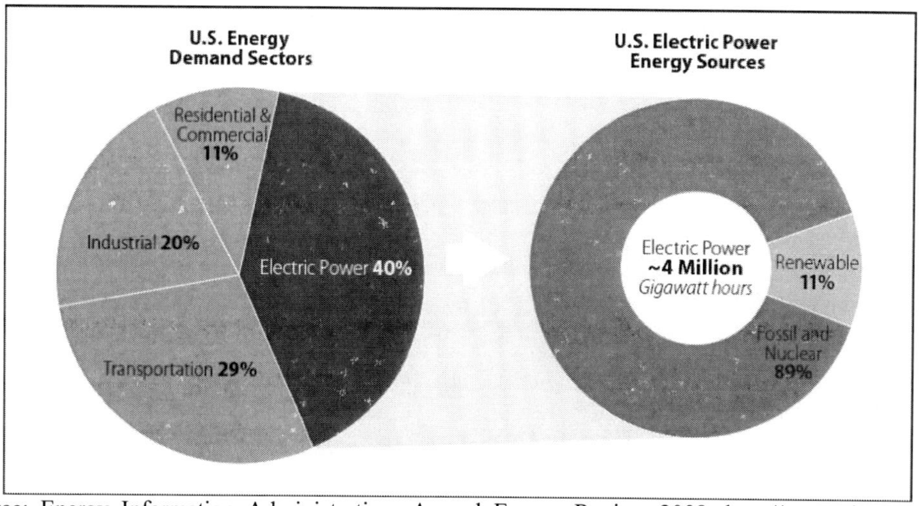

Source: Energy Information Administration, Annual Energy Review 2009, http://www.eia.doe.gov/ totalenergy data/annual/pdf/pecss_diagram_2009.pdf.

Figure 2. Supply Sources for U.S. Electric Power Sector.

These concerns are counter-balanced by the fact that fossil fuel electricity generation has long been—and generally continues to be—the least expensive form of electricity generation, by the fact that the United States has access to considerable resources of coal and natural gas for electricity generation, and from the economic and cultural inertia of the existing infrastructure in place for coal and natural gas to be used in large quantities for electricity generation. Renewable electricity generation provides two advantages when compared to fossil generation: (1) it relies on energy sources that may not decline over time, and (2) it produces little or no net greenhouse gas emissions or other pollutants during use.[3] However, renewable electricity generation does have liabilities and implementation challenges that will be further discussed in this report.

This report addresses two fundamental questions about U.S. renewable electricity generation potential: (1) How much renewable electricity generation might be possible in the United States?[4] and (2) What technical, operational, and economic challenges might renewables encounter when considering large-scale deployment for electricity generation?

Renewable Electricity Concepts and Units

Definition and Characteristics of Renewable Electricity

Renewable electricity is derived from renewable energy sources that "regenerate and can be sustained indefinitely."[5] This report does not use the term "clean energy" or "alternative energy," which are terms used by some to include renewable energy resources *plus* other sources that may emit little or no carbon dioxide during use, such as nuclear plants and coal-fired power plants equipped with carbon capture and sequestration capabilities. A discussion of biomass used to generate electricity is included, but biofuels are mentioned only briefly. This study is focused on wind, solar, hydroelectric, geothermal, biomass, and ocean/hydrokinetic energy sources used to generate electricity.

"Renewable energy" sources for electricity generation are often discussed as if they were a single entity, but renewable energy sources are more numerous and variable than fossil energy sources. Fossil fuels comprise oil, natural gas, and coal. The three major types of fossil fuels are extracted from the earth's crust by drilling or mining. Each of these fuels has very high energy density and is used primarily through combustion to exploit the heat produced. Renewable energy sources are more numerous and diverse and, thus, harnessing renewable energy requires a number of different technologies. Some of the distinctive characteristics of renewable energy are:

- *Renewable energy sources for electricity generation are numerous.* Sun, wind, flowing water in streams, flowing water in tidal channels, wave action in oceans, the earth's natural heat, biological materials, and others comprise the current portfolio of renewable energy sources, and additional renewable sources may be identified in the future.
- Each renewable energy source may be exploited in multiple ways to generate electricity using different technologies and materials. For example, the energy of the sun may be used by concentrating the energy to generate steam that drives electric

turbines (concentrating solar power), or the energy of the sun may be converted directly to electricity using semiconducting materials (photovoltaics). Furthermore, photovoltaic electricity may be produced using solar panels that consist of crystalline silicon, cadmium telluride, or other materials, and each material has unique characteristics.

- Renewable energy sources for electricity generation are naturally dispersed with relatively low energy densities. Fossil energy sources are typically concentrated as liquids or solids by millions of years of natural heating and pressure processes, which result in relatively high energy density that is accessible in wells or mines. In contrast, renewable energy sources are typically diffuse and require multiple technologies and management systems to gather and concentrate the resources.
- Each renewable electricity generation project/installation can vary in size. Renewable electricity generation systems are being installed in large, megawatt-scale projects that feed electricity into the electric grid for consumption along with electricity from other sources. In fact, the largest electric power plant in the United States is a hydroelectric facility, Grand Coulee Dam, which has a capacity of 7.08 GW.[6] At the same time, individual homes are being powered by small, kilowatt-scale rooftop solar panels. Also, wind turbines may be large, up to 5 megawatt (MW) utility-scale turbines, or small, approximately 5 kilowatt (kW) residential scale units.

Renewable Electricity Terminology and Units

This section defines terms and units used to describe and quantify renewable energy sources, and electricity generation potential from these sources, and how renewable energy might be compared to other forms of energy. Although this report focuses on renewable energy, discussions of fossil fuel units and consumption are included to facilitate comparisons with renewable forms of energy.

Measuring Energy: Fossil versus Renewable

Fossil fuels have traditionally been measured and marketed in the units of the physical material— barrels (42 gallons) of oil, short tons (2,000 pounds) of coal, or cubic feet of natural gas— transported to the point of end use. The use of volume or weight for measuring fossil fuels makes it challenging to compare the energy content among fossil fuels, and also contributes to the difficulty in clearly communicating the amounts of renewable energy that will be needed to replace fossil fuels. Each fossil fuel unit of measure has a corresponding energy content, which is typically expressed in terms of British Thermal Units (Btu).[7]

With the exception of biomass (typically measured in tons), each renewable energy source has its own unit of measure that may not be expressed as volume or weight. For example, wind energy is typically expressed in terms of *wind speed* (reported as meters per second); solar energy is typically expressed in terms of daily *insolation* (reported as kilowatthours per meter-square per day); hydroelectric is derived from flowing water, typically expressed in terms of water *flow rate* and *velocity*. In order to estimate annual electricity generation potential from renewable energy sources, experts must make assumptions about conversion equipment efficiencies and annual hours of operation.

Expressing Renewable Electricity Generation Potential: Watthour

Renewable electricity generation potential is typically expressed in terms of watthours (see text box below). A watthour (Wh) is a unit of electrical energy that can be generated, distributed, and consumed. A watthour can also be purchased and/or sold. For example, a residential electricity bill is typically calculated by multiplying the number of kilowatthours (kWh) consumed by a residence times the rate per kilowatthour charged by the electric power provider.[8] In 2009, U.S. total electricity net generation was approximately 4 million gigawatthours.[9] For the purpose of this report, renewable electricity generation potential, for all renewable energy sources, is expressed in terms of annual gigawatthours (GWh).

Authoritative Data Sources for Renewable Energy Resources

Various renewable electricity resource estimates for the United States are calculated by different institutions that use different processes, methodologies, and assumptions. No uniform methodologies exist for estimating and comparing the resource potential of different forms of renewable energy that might be used to generate electricity (see text box below).

Power Versus Energy: What's the Difference between a Watt and a Watthour?

Some energy reports provide statistics in units of power while other reports use units of energy. Power and energy are related, but they are not the same thing. Energy equals power multiplied by the amount of time the power is applied. Conversely, power is the rate at which energy is produced or consumed. Power is measured in watts, energy is measured in watthours. An electrical generator with 50 megawatts of power (or nameplate capacity) would generate 50 megawatthours of electrical energy for each hour it operates. The power capacity of a generator conveys only the size of the device, thus, when it's not operating, a generator does not produce any energy even though the power capacity remains the same. Power capacity is a critical variable when selecting a device to do a specific job, but the energy produced by the device depends on the amount of time it operates. The report examines total energy production with little regard to the size of the devices that produce it.

Traditional fossil fuel energy resource assessments are conducted through detailed geologic studies and the application of rigorously vetted methodologies. In contrast, most renewable electricity generation resource estimates are subject to the unique methods and assumptions of the organization conducting the assessment. Fossil energy resource estimates typically classify resources into categories such as: resource base, technically recoverable, economically recoverable, and reserves.[10] In principle, renewable energy resources should be measurable using similar analysis of the natural processes (wind, solar insolation, water flow, geothermal heat, etc.), adjusted for the effectiveness of the respective energy extraction technologies, and then couched in economic terms based on economic conditions and parameters.

In reality, it is very difficult, time consuming, and expensive to collect high quality data for wind, solar, stream flow, geothermal and biomass energy at a fine scale over the entire nation on an hourly, daily, or seasonal basis, as appropriate. In addition, the basic physics are different for extracting energy from solar, wind, hydro, geothermal, and biomass sources. Those physical differences give rise to different technologies, and many renewable energy technologies exhibit dramatically different performance according to geographic location and time of day or time of year. Therefore, estimating the amount of each type of renewable energy that is available to the nation is a challenging task. Examination of the literature reveals that estimates of available renewable energy resources vary widely. Attempting to compare estimates for different types of renewable energy multiplies those challenges.

The most reliable data for the various renewable energy sources come from national data collection programs from federal agencies. For example, the National Renewable Energy Laboratory, funded by the Department of Energy and its partners, operates programs designed to collect such data and has worked to identify the areas within the United States that are optimally suited for exploitation of various renewable energy sources. Other federal and state agencies, federal labs, and academic institutions also collect, analyze, and report renewable energy resource data. These data and estimates change over time as data collection technologies advance and understanding of the natural processes improves. Nevertheless, comparing renewable energy assessments from different sources is difficult, and a complete and comprehensive assessment of all available renewable energy resources for the nation does not yet exist. Collection of high quality data on renewable energy sources at a fine scale over broad ranges of time and geography will likely be an ongoing need for the nation. *Table 1* summarizes sources of information reviewed for this report.

Table 1. U.S. Renewable Electricity Generation Potential—Information Sources

Renewable Electricity Resource	Sources of Data Reviewed
Wind	National Renewable Energy Laboratory (NREL); American Wind Energy Association (AWEA); Department of Energy (DOE) Office of Energy Efficiency and Renewable Energy (EERE).
Solar	NREL; Solar Energy Industries Association (SEIA); DOE EERE.
Hydro	DOE EERE, Oak Ridge National Laboratory (ORNL); Idaho National Laboratory (INL); National Hydropower Association (NHA).
Geothermal	United States Geological Survey (USGS); NREL; Massachusetts Institute of Technology (MIT); Geothermal Energy Association (GEA).
Ocean-Hydrokinetic	Electric Power Research Institute (EPRI); DOE EERE; New York University; Ocean Renewable Energy Coalition (OREC).
Biomass	DOE EERE; United States Department of Agriculture (USDA); NREL, Biomass Power Association (BPA).

Source: CRS.

How Much Renewable Energy Is Available? It Depends ... and It Can Change

The overriding goal of this report is to provide Congress with accurate, comparable, and current U.S. renewable electricity resource estimates using currently available data. Answering the question "How much renewable electricity is possible in the United States?" is the primary objective. However, the answer to this key question is "it depends, and it will very likely change over time."

No centralized authoritative body or organization currently exists to develop and enforce standards for renewable electricity resource assessment methodologies and assumptions used to calculate estimates. Renewable electricity resource estimates come from multiple organizations. As a result, renewable electricity generation potential estimates are derived using different methodologies and different assumptions which, in turn, produce different estimates.

Estimates for renewable electricity generation potential in the United States depend on several factors such as the methodology used to calculate the estimates and certain assumptions that can have a major impact on the calculation results. With regard to methodologies used, resource estimates surveyed for this report came from several different organizations that include federal labs, industry organizations, and academic institutions. Each organization typically uses a unique methodology to calculate resource estimates. Therefore, comparing all of these estimates on an "apples-to-apples" equivalent basis is a challenge.

Key assumptions made for calculating renewable electricity generation potential can also have a major impact on resulting estimates. Geothermal electricity is a good example of how assumptions can impact renewable electricity generation estimates. Both USGS and MIT have published reports that estimate the amount of electricity generation potential from U.S. geothermal resources.[11] However, MIT estimates are more than 10 times larger than those from USGS. Two key assumptions explain most of this discrepancy: (1) Resource depth: The USGS geothermal study only considered geothermal potential at depths of 6 kilometers below the earth's surface whereas the MIT report considered depths of 10 kilometers, and (2) Which U.S. states were included: The USGS study only included 14 western states, Alaska, and Hawaii, while the MIT study included all 50 states. Thus, understanding assumptions for understanding and comparing the various resource potential estimates is critical.

Understanding certain exclusions for the various resource potential estimates is also important. Many of the studies surveyed for this report excluded certain areas from development based on several factors (national parks, urban areas, etc.). However, the types of exclusions and the constraints that result from exclusions vary. Comparing wind estimates and hydroelectricity estimates is one example. Wind electricity generation potential estimates exclude certain land areas. After these exclusions are taken into account, the NREL study referenced for this report assumes that wind projects can be built anywhere as long as the wind resource is large enough to meet certain electricity production levels. Hydroelectricity generation estimates, on the other hand, also include certain land area exclusions but apply additional filters such as the location being within one mile of a road and a transmission line.

If identical exclusions were applied to all renewable electricity generation resource assessments, resource potential results may be quite different. Further, estimates for renewable electricity generation potential in the U.S. will likely change over time as resource estimate methodologies improve, renewable electricity generation technologies are developed and commercialized, and better information about the magnitude and quality of renewable energy resources is made available. As a result, estimates of renewable electricity generation potential could either go up or down in the future.[12]

U.S. Renewable Electricity Use and Potential

This section provides a brief overview of current U.S. renewable electricity generation, followed by a series of discussions of specific renewable electricity technologies. Current electricity generation, estimated potential generation, and deployment challenges are discussed for wind, solar, geothermal, hydroelectric (hydro), ocean-hydrokinetic, and biomass energy sources.

Summary of Current U.S. Renewable Electricity

In 2009 renewable energy resources provided 11% of U.S. electricity net generation. Renewable electricity was derived from wind, solar, geothermal, hydro, and biomass energy sources. The largest source of renewable electricity was hydro. Wind and biomass each contributed between 1% and 2% of total U.S. electricity net generation. Solar and geothermal electricity generation contributed relatively small amounts to the renewable electricity portfolio mix (*Figure 3*).

(Percentage of each renewable source)

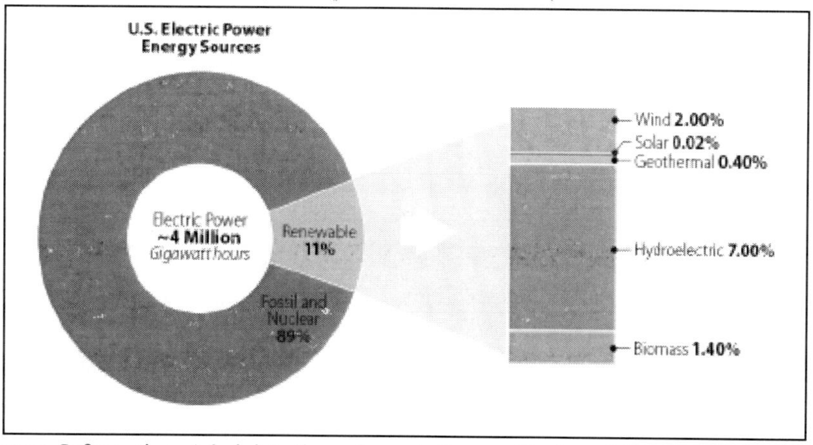

Source: Energy Information Administration, Annual Energy Review 2009, http://www.eia.doe.gov /totalenergy/ data/annual/pdf/pecss_diagram_2009.pdf.

Notes: Renewable electric power percentages may not add to 11% because of independent rounding error.

Figure 3. U.S. Electricity Generation from Various Renewable Sources, 2009.

Future Renewable Electricity Generation Potential

The following sections discuss the estimated range of electricity generation potential from wind, solar, geothermal, hydro, ocean-hydrokinetic, and biomass renewable energy sources. A discussion of technology and cost considerations is presented for each respective renewable source of electricity. As discussed above, comparing renewable electricity generation resource estimates is a challenging task. The approach used to derive the resource estimate range for each technology is described in the footnotes to each renewable energy source section. Table 2 provides a summary of U.S. renewable electricity generation potential, based on the research and analysis performed for this report, current and projected renewable electricity generation potential, cost of electricity estimates, and a summary of key challenges for each renewable energy source. As the table shows, renewable electricity generation potential is compared to 2009 total U.S. net generation of approximately 4 million gigawatthours (GWh). This approach was used in order to indicate the maximum electricity generation contribution that might be available from each renewable energy source. Furthermore, the reader is advised that levelized cost of energy (LCOE) estimates presented in *Table 2* only reflect electricity costs associated with capacity additions in EIA's Annual Energy Outlook 2011. For more information, see the "Levelized Cost of Energy (LCOE)" section below.

Research and analysis conducted for this report indicates that renewable energy sources may, theoretically, have the potential to satisfy a large portion of U.S. electric power needs. However, numerous technical, operational, economic, and practical challenges will likely be encountered, which may ultimately limit the potential contribution of renewable electricity generation. These challenges are discussed in a following section. Furthermore, while the potential for renewable electricity generation in the country is vast, EIA Annual Energy Outlook 2011 reference case projections indicate that renewables will contribute between 14% and 15% of total U.S. electricity generation by 2035.[13] Also, the quality of resources estimates is different for each renewable technology, and these estimates may change as new data are collected and new assessments are conducted. The current estimates represent a snapshot in time and must be continually updated as additional data become available.

Wind

U.S. Resource Estimates

U.S. wind energy resource estimates are highly dependent on certain assumptions used to calculate them, and users of those estimates should pay careful attention to the underlying assumptions. Turbine height and capacity factor assumptions can have major impacts on wind resource estimates. For example, winds are generally stronger at greater heights above the ground. As a result, wind resource estimates at 100 meters are likely to be greater than those at 50 meters.

Table 2. Summary of U.S. Renewable Electricity Resources and Challenges

Electricity Generation Potential	Wind[a]		Solar[b]		Geothermal		Hydro		Ocean-Hydrokinetic		Biomass[c]	
	Low	High	Low	High	Low	High	Low	High	Low	High	Low	High
Electricity Potential (GWh/yr)	32,500,000	61,400,000	4,000,000	56,300,000	927,791	36,991,864	558,145	613,333	287,850	2,161,350	125,730	1,428,780
* Total estimated resource potential												
% of 2009 total U.S. generation	>100%	>100%	100%	>100%	23%	>100%	14%	16%	7%	55%	3%	36%
Current and Forecasted Generation												
2009 Generation (GWh)	73,886		891		15,009		273,445		not available		54,493	
% of 2009 total U.S. generation	1.87%		0.02%		0.38%		6.92%		not available		1.38%	
EIA LCOE[d] $/MWh ** Only for capacity additions forecasted in AEO 2011	$82 to $349		$159 to $642		$92 to $116		$59 to $121		not available		$99 to $134	
2035 Generation (GWh) *** EIA AEO 2011 forecast (reference case)	160,880		3,970		49,190		310,590		not available		47,440	
% of 2035 total est. generation	3.48%		0.09%		1.06%		6.70%		not available		1.02%	
Deployment Challenges, Issues, and Barriers?												
Power System Integration	Yes		Yes		No		No		Yes		No	
Transmission	Yes		Yes		Yes		Yes		Yes		Yes	

Table 2. (Continued)

Electricity Generation Potential	Wind[a]	Solar[b]	Geothermal	Hydro	Ocean-Hydrokinetic	Biomass[c]
	Low High	Low High	Low High	Low High	Low High	Low High
Cost	No/Yes - Onshore wind costs are in the range of fossil electricity costs - Offshore costs are higher	Yes - Currently the highest cost source of renewable electricity	Yes - Enhanced geothermal system (EGS) costs are estimates only; NREL indicates EGS LCOE could be as high as $1,000/MWh	No/Yes/ -One of the oldest and lowest-cost sources of renewable electricity - Emerging small/low-head hydro costs unknown	Yes - Actual cost of ocean and hydrokinetic electricity is unknown	Yes - Cost of electricity can be impacted by logistics and feedstock quality
Intermittency/ Variability	Yes - Wind resources can vary on an hourly, daily, and/or annual basis	Yes - Cloud coverage and other weather events can degrade solar technology performance; solar energy not available at night	No - Geothermal electricity production can be predictable and may operate at high capacity factors	Yes - Hydropower resource can vary based on annual rain/snow fall	Yes/No - Wave energy resources can vary based on the amount of wind; tidal energy may be predictable	No - Biomass electricity plants can operate at high capacity factors and may provide baseload power
Technology	No/Yes - Onshore wind technology considered commercial - Offshore wind may require further technology development to operate in harsh ocean environment	Yes - Emerging PV technologies that may improve efficiencies & reduce costs - Some CSP technologies may require further engineering, development, and demonstration before being commercial	Yes - EGS, the largest potential source of geothermal electricity, technology is not yet commercially available - Specialized drilling equipment may be required	Yes - Small and low-head/lowpower technologies are being developed and matured but some are not yet commercially available	Yes - Ocean and hydrokinetic energy technologies are considered "emerging" with no commercially available electricity generation technologies	No/Yes - Biomass combustion technology might be considered commercial - Some technical issues (tar production, equipment fouling etc.) may have to be addressed
Environmental Impact	Yes - Land use, habitat and scenic disturbance,	Yes - Water use requirements for some	Yes - Water use; discharge of metals and toxic gas	Yes - Ecosystem changes; fish	Yes - Alteration of currents and waves;	Yes- Biomass combustion for electricity

Electricity Generation Potential	Wind[a]	Solar[b]	Geothermal	Hydro	Ocean-Hydrokinetic	Biomass[c]
	Low High	Low High	Low High	Low High	Low High	Low High
	noise, and bird mortality are potential	solar thermal technologies	- Ground/surface water	migration and mortality - Habitat	alteration of sediment disposition;	generation emits
	environmental issues associated with wind projects	- Land use and associated habitat disturbance - Mobilization of trace metals	pollution - Land subsidence and seismicity	damage; water quality degradation	habitat impacts; noise; electromagnetic fields; toxicity of lubricants and other fluids; animal injury from moving parts; degradation of water quality	NOx, CO_2, and other emissions - Land use/change associated with biomass production - Carbon neutrality of biomass combustion is a possible issue
Infrastructure	Yes - Offshore wind may require specialized vessels, portside infrastructure, under-sea transmission, etc.	Unknown - Information regarding potential infrastructure issues not available	Unknown -Information regarding potential infrastructure issues not available	Unknown - Information regarding potential infrastructure issues not available	Yes - Specialized infrastructure may be needed to install and maintain operational projects	Yes - Logistics infrastructure may be needed to gather and process biomass material
Materials and Resources	Yes - High volumes of steel, concrete, and rare-earth metals may be needed to support large scale wind deployment	Yes - High volumes of steel and concrete may be needed to support large-scale deployment of utility -scale solar; silicon, tellurium, cadmium, silver, and other commodities may be required for large-scale PV deployment	Unknown - Information regarding potential materials and resources issues not available	Unknown - Information regarding potential materials and resources issues not available	Yes - Materials that can operate for long periods of time in a corrosive ocean environment may need to be developed	Yes - Economical electricity generation from biomass may require adequate biomass resources within a defined geographic area

Source: CRS; Various sources as identified and referenced in the respective sections of this report.

[a] Includes both onshore and offshore wind; [b] Includes both photovoltaic and concentrating solar; [c] Does not include liquid biofuels used for transportation.

[d] LCOE = Levelized Cost of Energy.

Wind energy resources in the United States are typically categorized as either "onshore" or "offshore." According to National Renewable Energy Laboratory (NREL) estimates, onshore wind electricity generation potential for the 48 contiguous United States ranges from 22.5 million gigawatthours to 46.9 million gigawatthours annually.[14] Wind resources can vary state by state and region by region. Based on NREL estimates, the largest onshore U.S. wind energy resources are located in the middle of the country (see *Figure 4).*

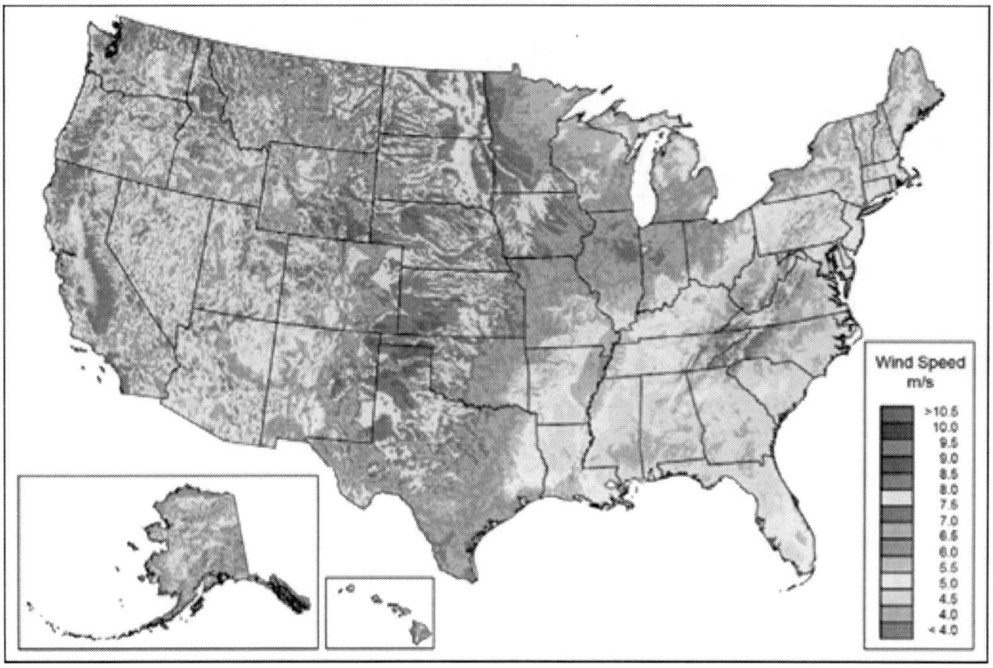

Source: http://www.windpoweringamerica.gov/wind_maps.asp.

Figure 4. U.S. Onshore Wind Energy Resources, 80 Meter Turbine Height.

NREL has also calculated estimates for U.S. offshore wind energy resources.[15] Based on NREL estimates, offshore wind energy resource potential may range between 10 million GWh and 14.5 million GWh annually.[16] *Figure 5* illustrates how offshore wind energy resources vary by location. However, in its February 2011 National Offshore Wind Strategy report, the DOE EERE states that "the offshore wind resource is not well characterized."[17] This uncertainty indicates that additional work may be needed to more accurately assess U.S. offshore wind potential.

Technology and Cost Considerations

Onshore wind energy conversion technology is generally considered to be commercially available[18] and many projects are able to attract debt and equity investment capital for project development. General Electric and Siemens were the top two manufacturers of wind turbines installed in the U.S. during 2010.[19] Typical wind turbines have a rated capacity between 1 megawatt and 3 megawatts, and the general industry trend is to continue increasing the size and capacity of individual wind turbines in order to operate at greater heights (taller towers) and realize economies of scale by generating more watthours from a single unit. Offshore wind energy technology faces some technical challenges associated with operating in a corrosive marine environment and installation of equipment at various water depths. The Department of Energy (DOE) operates a Wind Power Program aimed at wind research and development needs.[20]

The cost of wind-generated electricity can vary based on a number of technology, performance, operational, and financial factors. These factors are discussed in the "Levelized Cost of Energy" section below. Assumptions made for these factors can result in significant differences among cost-of-electricity estimates. In its Annual Energy Outlook 2011, EIA estimates onshore wind electricity costs to range from $82-$115 per megawatthour (MWh) and offshore electricity costs between $187-$349 per MWh.[21] *Figure 12* provides a comparison of costs for conventional (fossil and nuclear) and renewable electricity generation.

(Excluding Alaska)

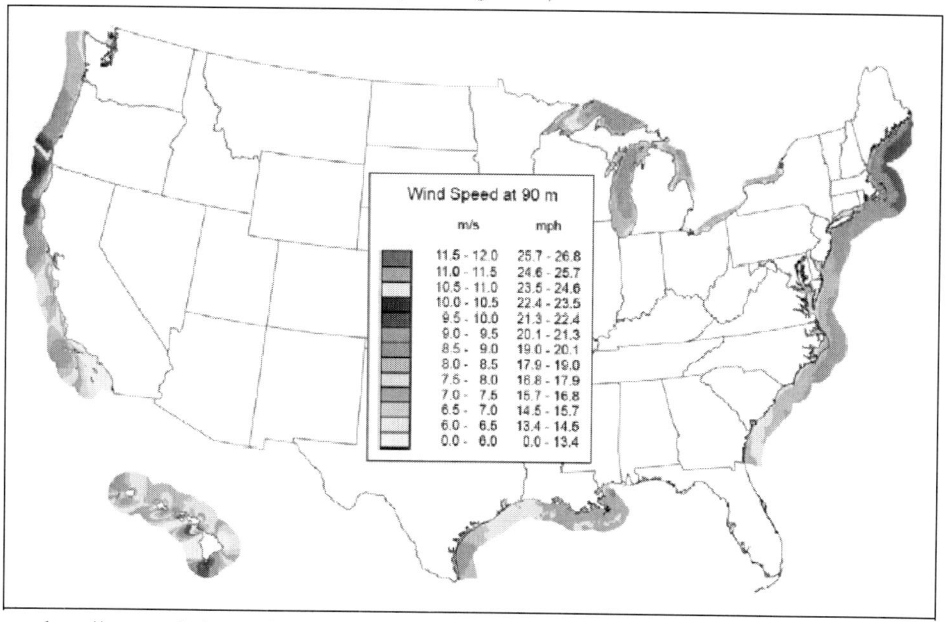

Source: http://www.windpoweringamerica.gov/windmaps/offshore.asp.

Figure 5. U.S. Offshore Wind Energy Resources, 90 Meter Turbine Height.

Solar

U.S. Resource Estimates

Every U.S. locale receives sunlight during a calendar year, of course, but the amount of radiation that reaches a given point at a particular time can vary based on factors that might include geography (including latitude), time of day, season, landscape, and weather.[22] Two authoritative sources for U.S. solar resources are the National Solar Radiation Database (NSRDB), and NREL and State University of New York at Albany (SUNYA) satellite-derived solar resources.[23]

Quantifying solar resource data, in terms of annual electricity generation potential, is complicated by several factors.[24] First, two different methods are used for capturing and converting solar energy into electricity: (1) concentrating solar power (CSP), and (2) photovoltaic (PV) solar power.[25] Second, solar radiation has different components that may be better suited for different collector types.[26] Third, different system configurations are used to collect data and calculate solar resource estimates.[27] Finally, different types of CSP and PV technologies, with different cost, efficiency, and performance characteristics, are available for solar energy conversion.[28]

CSP resources vary throughout the country, with most of the highest quality resource located in the southwestern United States (see *Figure 6*). CSP electricity generation is typically better suited for large-scale (greater than 10 MW) power generation projects. NREL and Sandia National Laboratories estimate that U.S. CSP electricity generation potential is approximately 16.3 million GWh.[29] This estimate is the result of applying a set of filters to existing CSP resource data in order to calculate CSP electricity generation potential.[30] NREL and Sandia CSP estimates include electricity generation potential in seven U.S. states.[31]

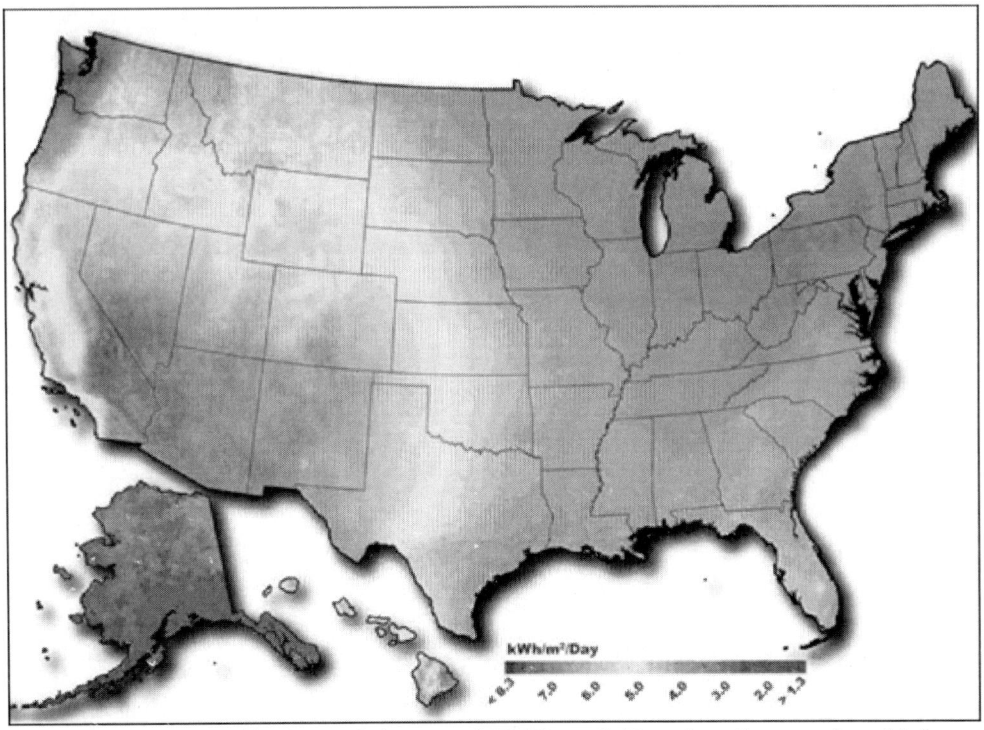

Source: National Renewable Energy Laboratory (NREL), available at http://www.nrel.gov/gis/images map_csp_national_lo-res.jpg.

Notes: Annual average direct normal solar resource data are shown. The data for Hawaii and the 48 contiguous states are 10km satellite modeled dataset (SUNY/NREL, 2007) representing data from 1998-2005. The data for Alaska are a 40 km dataset produced by the Climatological Solar Radiation Model (NREL, 2003); kWh/m2/Day = kilowatthour per square meter per day.

Figure 6. U.S. Concentrating Solar Resource.

Photovoltaic resources also vary throughout the country and, much like CSP, the highest quality PV resources are located in the southwestern United States (see *Figure 7*). PV systems offer flexibility in terms of project size and can be used for residential, commercial, and utility-scale applications.

As of July 2011, estimates of the technically available and economically recoverable solar photovoltaic resource, in terms of annual GWh, were not available. However, previous work by NREL provides some indication about solar generation potential. NREL researchers analyzed land-use requirements for generating 100% of U.S. electricity and estimated that 0.6% of total U.S. land area would be needed to satisfy current demand load using a "base system configuration."[32] NREL has also estimated land-use requirements for a variety of other system configurations.[33]

Calculating total PV generation based on this NREL analysis is somewhat complicated, but it may be reasonable to assume that solar PV could theoretically generate 10 times the amount of current U.S. demand, although realizing this amount of electricity generation may be limited by several factors, particularly cost and power system integration.[34] Extrapolating from the NREL analysis, U.S. annual solar PV generation potential may be equal to approximately 40 million GWh. NREL has also evaluated the generation potential of residential and commercial rooftop PV systems and estimates that, under a base-case scenario, approximately 819,000 GWh of electricity could be generated each year using existing rooftop space.[35]

Technology and Cost Considerations

The most commonly used CSP technology in the United States is the parabolic trough. Of the 509 megawatts of U.S. installed CSP capacity, approximately 98% uses parabolic trough technology.[36] Crystalline silicon is the most commonly used PV technology.[37] While some CSP and PV technologies might be considered commercially available, there are a number of research and development activities within CSP and PV markets.[38] Generally speaking, most CSP and PV R&D work is focused on improving system-level efficiencies and reducing system costs. Storage, demand response, and other "smart-grid" technologies may further enable large-scale solar deployment.[39]

The cost of solar electricity has been a challenge faced by both CSP and PV technologies. Solar electricity, according to the Energy Information Administration, is the highest-cost source of electricity generation, with CSP costs ranging from $192-$642 per MWh and PV costs ranging from $159-$324 per MWh.[40] DOE is funding an initiative, known as the SunShot program, which aims to reduce the cost of PV electricity generation to $60 per MWh.[41] *Figure 12* provides a comparison of costs for conventional (fossil and nuclear) and renewable electricity generation.

Geothermal

U.S. Resource Estimates

Geothermal energy is present throughout the entire country, with most of the highest-quality geothermal resources generally located in the western United States, Alaska, and Hawaii.[42]

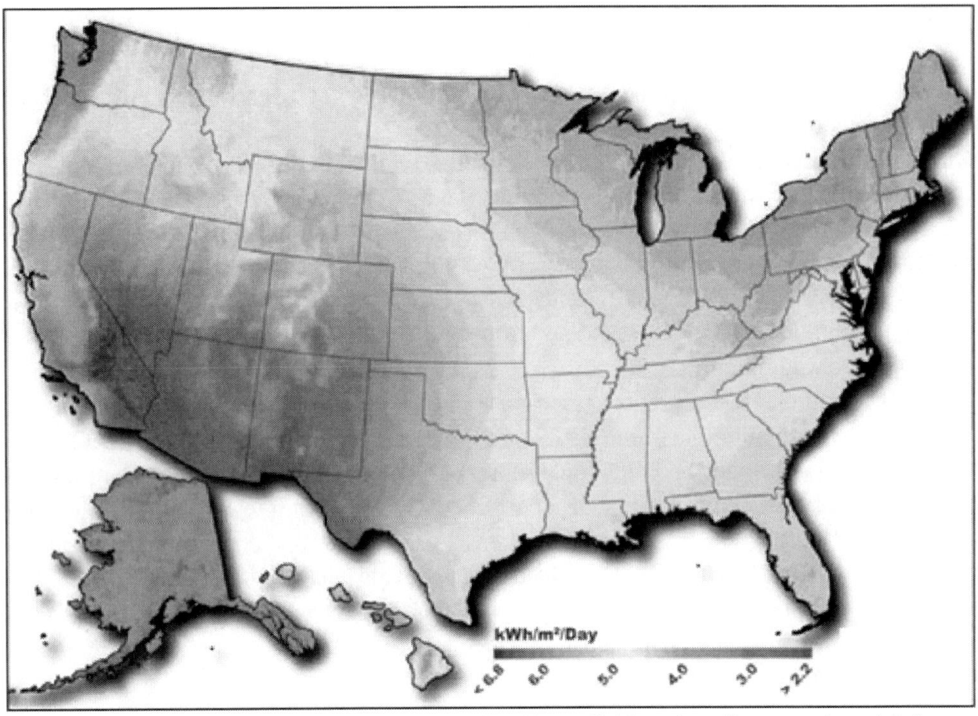

Source: National Renewable Energy Laboratory (NREL), available at http://www.nrel.gov/gis/images/map_pv_national_lo-res.jpg.

Notes: Annual average solar resource data are shown for a tilt-latitude collector. The data for Hawaii and the 48 contiguous states are a 10 km satellite modeled dataset (SUNY/NREL, 2007) representing data from 1998- 2005. The data for Alaska are a 40 km dataset produced by the Climatological Solar Radiation Model (NREL, 2003); kWh/m2/Day = kilowatthour per square meter per day.

Figure 7. U.S. Photovoltaic Solar Resource.

However, all states may have geothermal electricity generation potential through the use of enhanced, or engineered, geothermal systems (EGS) technology.[43] USGS, NREL, and the Massachusetts Institute of Technology (MIT) each have published estimates for U.S. geothermal electricity generation. Generally, geothermal electricity generation resources are classified into three categories: (1) identified resources, (2) undiscovered resources, and (3) enhanced geothermal systems.[44] *Table 3* provides a summary of USGS, NREL, and MIT potential geothermal capacity estimates for these resource types along with annual electricity generation potential in GWh.

USGS, NREL, and MIT each have a different estimate for U.S. geothermal electricity generation potential, especially with regard to enhanced geothermal systems. Two primary factors account for the differences in estimates: (1) USGS estimates are confined to western U.S. states, Hawaii, and Alaska, while NREL and MIT estimates include potential electricity generation from all 50 states, and (2) USGS estimates are for resource depths between 3 kilometers (km) and 6 km, while NREL and MIT estimates are for resource depths between 3 km and 10 km.[45] This disparity in resource estimates illustrates how assumptions can significantly alter the assessment results.

Table 3. U.S. Geothermal Electricity Generation Potential

	Identified Resource		Undiscovered Resource		Enhanced Geothermal Systems	
	Low	High	Low	High	Low	High
USGS[a] Capacity (MW-e)	3,675	16,457	7,917	73,286	345,100	727,900
Electricity Generation[b] (GWh/yr)	29,618	132,630	63,805	590,627	834,369	1,759,888
NREL[c] Capacity (MW-e)	n/a	6,390	n/a	30,030	n/a	15,000,913
Electricity Generation[b] (GWh/yr)	n/a	51,498	n/a	242,018	n/a	36,268,607
MIT[d]						
Capacity (MW-e)	n/a	n/a	n/a	n/a	1,249,000	12,486,000
Electricity Generation[b] (GWh/yr)	n/a	n/a	n/a	n/a	3,019,782	30,188,151

Source: U.S. Geological Survey, National Renewable Energy Laboratory, MIT. See specific references below.

Notes: Electricity generation potential estimates represent what might be technically recoverable and do not include any filters for economic factors. "Identified" and "Undiscovered" resources generally represent conventional geothermal resources where naturally occurring conditions (high temperature and permeability) allow for extraction of geothermal energy. "Enhanced Geothermal Systems" require engineering of rock permeability to create geothermal energy extraction conditions. Electricity generation numbers were calculated by CRS using a 92% capacity factor for each geothermal capacity estimate. USGS low estimates for each resource category represent resources that have a 95% probability of being available. USGS high estimates for each resource category represent resources that have a 5% probability of being available. NREL analysis provided a single number for identified, undiscovered, and EGS resource estimates, respectively. The MIT study focused on the potential of EGS in the United States. The large difference between MIT's low and high estimates reflect an assumption made for the EGS energy recovery factor (2% for the low estimate, 20% for the high estimate).

[a] "Assessment of Moderate- and High-Temperature Geothermal Resources of the United States," U.S. Geological Survey, 2008, available at http://pubs.usgs.gov/ fs/2008/3082/pdf/fs2008-3082.pdf.

[b] Annual electricity generation potential assumes that all potential geothermal resources are developed and operating. This is highly unlikely since EGS systems may result in resource depletion over a 30-40 year operating life; regeneration of this resource is estimated to take approximately 100 years. As a result, EGS estimates for GWh/yr were discounted by a factor of 0.3 in order to calculate sustainable electricity generation potential based on a 30-year depletion and 100-year regeneration profile.

[c]. Augustine, K. Young, and A. Anderson, "Updated U.S. Geothermal Supply Curve," National Renewable Energy Laboratory, Conference Paper presented at Stanford Geothermal Workshop, February 1, 2010, available at http://www.nrel.gov/docs/ fy07osti/41073.pdf.

[d] "The Future of Geothermal Energy: Impact of Enhanced Geothermal Systems (EGS) on the United States in the 21st Century," Massachusetts Institute of Technology, 2006, available at http://geothermal.inel.gov/ publications/future_of_geothermal_ energy.pdf.

MW-e = megawatt electrical generating capacity.

GWh/yr = gigawatthours per year.

Figure 8 shows conventional geothermal sites and the estimated relative suitability of EGS geothermal energy recovery throughout the U.S.

(Locations of identified hydrothermal sites and favorability of deep enhanced geothermal systems [EGS])

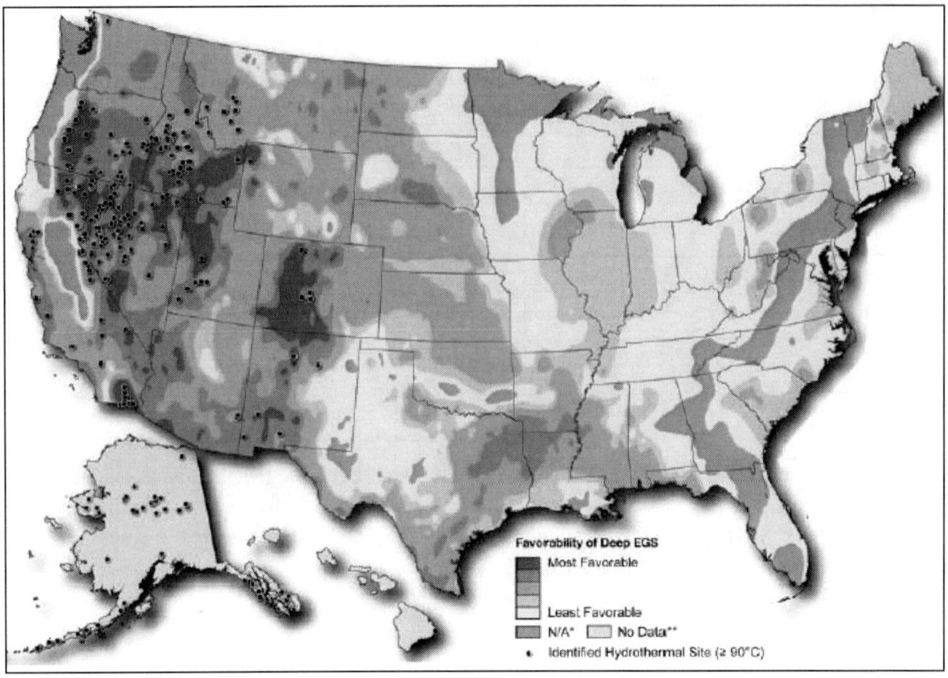

Source: NREL, available at http://www.nrel.gov/gis/images

Notes: Map does not include shallow EGS resources located near hydrothermal sites or USGS assessment of undiscovered hydrothermal resources. Source data for deep EGS includes temperature at depth from 3 to 10 km provided by Southern Methodist University Geothermal Laboratory (Blackwell & Richards, 2009) and analysis (for regions with temperatures ≥150«C) performed by NREL (2009). Source data identified hydrothermal sites from USGS Assessment of Moderate- and High-Temperature Geothermal Resources of the United States (2008).

* "N/A" regions have temperatures less than 150«C at 10 km depth and were not assessed for deep EGS potential.

** Temperature at depth data for deep EGS in Alaska and Hawaii not available.

Figure 8. Geothermal Resource of the United States.

Technology and Cost Considerations

For conventional hydrothermal geothermal resources, four commercially available technologies are available for generating electricity: (1) flash power plants, (2) dry steam power plants, (3) binary power plants, and (4) flash/binary combined cycle.[46] As of April 2011, U.S. geothermal installed capacity was 3,102 MW, which represents approximately 0.3% of total U.S. electricity capacity.

In 2009, 15,009 GWh of electricity was generated from geothermal energy sources. The majority of existing geothermal capacity is located in California.[47] Enhanced Geothermal Systems (EGS) technology could potentially enable large-scale deployment of economically recoverable geothermal electricity generation.[48] However, EGS technology has not been demonstrated at scale and is not yet commercially available.[49]

Several factors can influence the cost of geothermal electricity. These factors include the resource quality (temperature and volume), resource depth, drilling costs, and geothermal equipment costs.

EIA estimates that the levelized cost of energy (LCOE) for conventional geothermal electricity ranges from $92/MWh to $116/MWh.[50] *Figure 12* provides a comparison of costs for conventional (fossil and nuclear) and renewable electricity generation. NREL has also estimated geothermal electricity LCOE and concluded that, depending on the total amount of capacity installed, geothermal (conventional and EGS) electricity costs could range between $50/MWh and $1,200/MWh (2008 US$).[51] Based on these cost of energy estimates, NREL indicates that the amount of EGS resource "that can be economically produced is likely much smaller" than the total resource potential.[52]

Hydroelectric

U.S. Resource Estimates

Hydropower is currently the largest source of renewable electricity production in the United States. In 2010, approximately 257,000 GWh was generated from hydropower resources, equal to roughly 7% of total U.S. electricity generation.[53] Hydropower can be generated in many ways. For the purpose of this report, hydropower refers to "conventional" hydropower[54] and does not include hydrokinetic energy, ocean energy, or pumped storage.[55]

Since 1998, several Idaho National Laboratory (INL) reports have estimated the potential to develop new hydropower generation capacity. Three of those reports trace a time-wise increase in the estimates of potential generation capacity: 30 GW (1998),[56] 43 GW (2003),[57] and 60 GW (2006).[58]

A review of the INL reports revealed that estimates differed with regard to the hydropower categories included in the calculations.[59] After sorting through the studies and attempting to remove duplicative and non-relevant data, CRS calculated additional hydropower potential to be approximately 65 gigawatts, which equates to approximately 284,700 GWh of additional annual electricity generation potential.[60]

A 2007 report by the Electric Power Research Institute (EPRI) estimated that additional hydropower capacity potential was equal to 62.3 gigawatts.[61]

However, recently published Oak Ridge National Laboratory (ORNL) resource potential estimates for non-powered dams may increase the total hydropower resource assessment by as much as 12.6 gigawatts.[62]

Figure 9 provides summary information about the location of existing and potential hydroelectricity facilities in the United States.

Technology and Cost Considerations

Hydroelectricity generation in the United States dates back to the 1880s and many technologies are fully commercialized with proven operational performance.[63] However, DOE is pursuing efforts to further improve the performance, economics, and environmental impact of conventional hydropower technologies.[64]

Low-head and low-power hydroelectricity technology that might be used in constructed waterways, such as canals, may require additional research, development, and demonstration before being commercially available.[65]

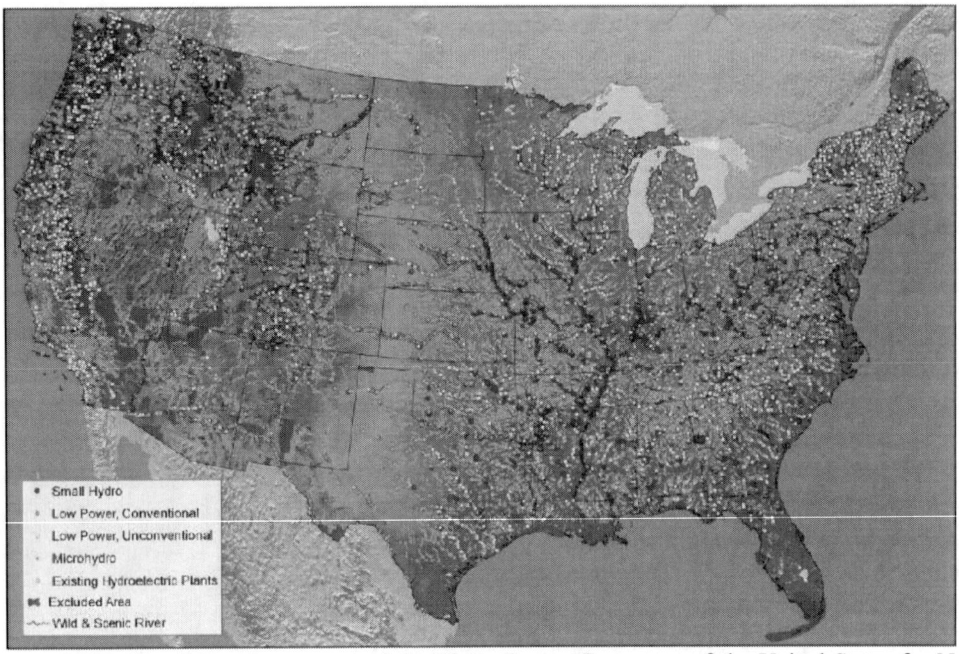

Source: DOE. "Feasibility Assessment of the Water Energy Resources of the United States for New Low Power and Small Hydro Classes of Hydroelectric Plants," DOE, Office of Energy Efficiency and Renewable Energy, Wind and Hydropower Technologies, January 2006, available at http://www1.eere.energy.gov/ windandhydro/pdfs/doewater-11263.pdf.

Notes: Alaska and Hawaii were included in the DOE study, but were not included in the accompanying map. DOE study results indicate that Alaska may have the potential to increase its hydropower capacity by as much as 16 times and Hawaii was distinguished as the state having the highest concentration (measured as kilowatt-annual per square mile) of hydropower potential.

Figure 9. Existing and Potential Hydropower Projects in the Lower 48 United States.

Hydroelectricity is generally considered to be one of the lowest-cost sources of renewable electricity. EIA estimates that the LCOE for new hydroelectricity plants ranges from $59/MWh and $121/MWh.[66] *Figure 12* provides a comparison of costs for conventional (fossil and nuclear) and renewable electricity generation. However, due to their relatively early stage of development, the cost of electricity from low-head and low-power technologies remains somewhat uncertain.

Ocean and Hydrokinetic

U.S. Resource Estimates

Ocean-based energy resources come in several forms, including (1) tidal, (2) wave, (3) current, and (4) thermal (also known as Ocean Thermal Energy Conversion or OTEC).[67] Each ocean energy resource is fundamentally different in terms of the amount of available resources, location of the resource, and the conversion technology used to generate electricity. While a limited number of ocean energy resource assessments are available, the Electric Power Research Institute (EPRI) published resource estimates for wave and tidal energy in

2006.[68] DOE has funded several resource assessments that are not yet available.[69] *Table 4* summarizes some of the resource estimates for different categories of ocean energy.

Table 4. U.S. Ocean Energy Resource Estimates

	Resource Estimate (GWh/year)		
	Low	High	Resource Assessment Status
Wave	255,000	2,100,000	In 2008, DOE awarded a Marine Energy Grant to EPRI to assess U.S. wave energy resources.
Tidal	n/a	6,600	Georgia Tech Research Corporation was awarded a grant from DOE to assess tidal stream energy production potential.
Current	n/a	n/a	DOE awarded a grant to Georgia Tech Research Corporation to create a database of ocean current energy potential.
OTEC	n/a	n/a	DOE awarded a grant to Lockheed Martin in 2009 to conduct global and domestic ocean thermal resource assessments.

Source: EPRI (Resource Estimates); "Report to Congress on Renewable Energy Resource Assessment Information for the United States," DOE, Office of Energy Efficiency and Renewable Energy, January 28, 2011.

Notes: n/a = not available.

Figure 10 illustrates the location and magnitude of U.S. wave energy resources. The majority of wave energy potential exists off the coasts of Alaska, Hawaii, and west coast states.

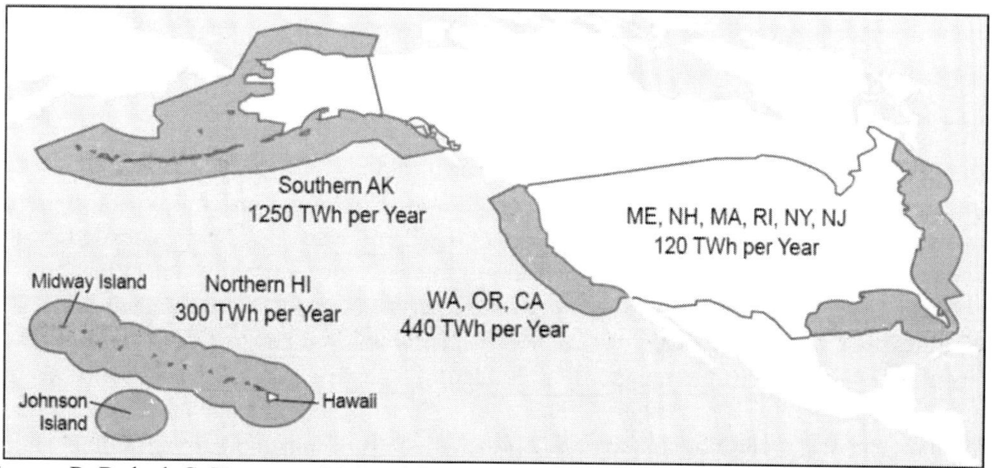

Source: R. Bedard, G. Hagerman, M. Previsic, O. Siddiqui, R. Thresher, and B. Ram, "Final Summary Report: Project Definition Study – Offshore Wave Power Feasibility Demonstration Project," Electric Power Research Institute, September 22, 2005, available at http://oceanenergy.epri.com/ attachments/wave/reports/ 009_ Final_Report_RB_Rev _2_092205. pdf.

Notes: TWh = terawatthours. 1 TWh = 1,000 gigawatthours.

Figure 10. U.S. Wave Energy Resources.

Hydrokinetic energy, defined as river in-stream energy for the purpose of this report, can be extracted from the natural water flow in rivers. The amount of electricity that can be generated from this energy source is dependent on the volume and velocity of the water resource. A DOE-funded study by New York University estimates that approximately 12.5 GW of hydrokinetic power potential might be possible.[70] Assuming a capacity factor between 30% and 50%, electricity generation potential from hydrokinetic resources may range from 32,850 GWh to 54,750 GWh.[71]

Technology and Cost Considerations

Ocean and hydrokinetic electricity generation technologies might be considered "emerging" as they have yet to operate at a significant commercial scale. Nevertheless, demonstration and commercial deployment of ocean and hydrokinetic projects is being pursued.[72] Many technology concepts are being developed and demonstrated, including more than 100 ocean energy devices worldwide, with approximately 30 under development in the United States.[73]

Given the early developmental status of ocean and hydrokinetic electricity production technologies, estimating the levelized cost of energy is challenging.[74] EIA did not include an LCOE estimate for ocean and hydrokinetic electricity generation as part of the Annual Energy Outlook (AEO) 2011.

Biomass

U.S. Resource Estimates

Accurate estimates for biomass electricity generation potential are somewhat challenging because biomass material (forest, agriculture, solid waste, and landfill gases) can be used in a variety of competing ways to include electricity generation, biofuel production, and space heating for residential and commercial buildings.[75] Also, unlike other renewable energy sources, biomass might be considered a managed resource in that the quantity of biomass material available for electricity generation can go up or down based on changes in management practices.[76] As a result, U.S. biomass electricity generation potential is highly dependent on how much biomass is available and how much biomass material is dedicated for this specific use. In 2009 an estimated 54,493 GWh of electricity was generated from biomass, which represented approximately 1.2% of total U.S. net electricity generation.[77]

According to DOE, approximately 190 million tons of biomass are consumed each year, with roughly 25% to 35% of current biomass consumption being used for electricity generation. DOE analysis and reports indicate that the potential may exist to produce about 1.3 billion tons of biomass annually.[78] However, estimating the amount of electricity that might be generated from biomass depends on the amount of biomass material available and on the portion of that material that might be used for electricity generation. *Table 5* provides an estimate for potential generation if DOE's 1.3 billion ton estimate of biomass production were realized.

Figure 11 indicates the relative concentration of current biomass resources throughout the United States.

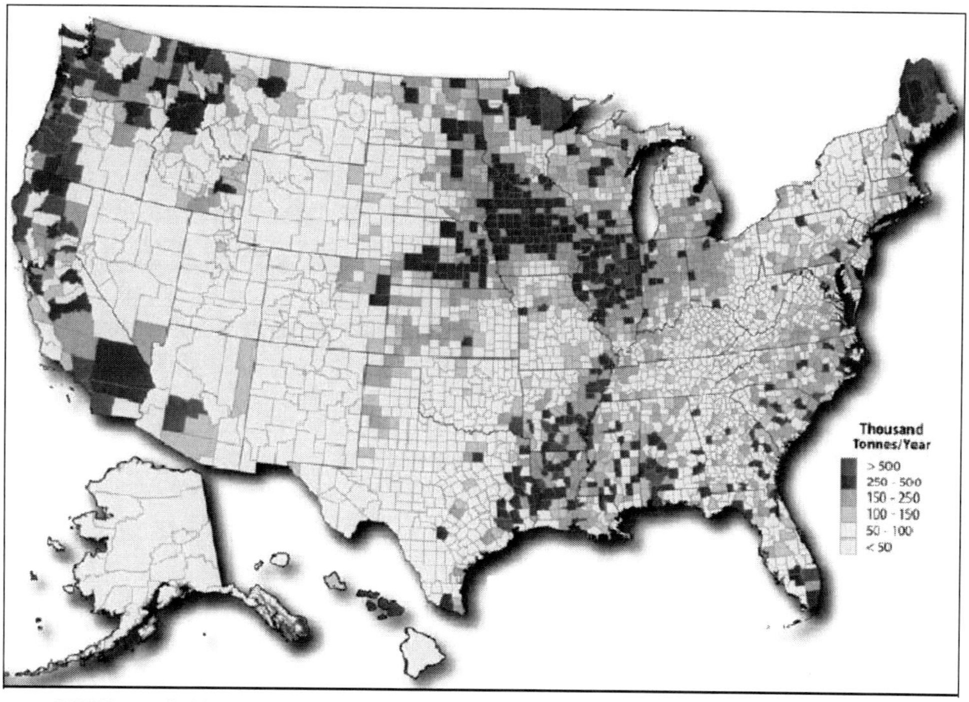

Source: NREL, available at http://www.nrel.gov/gis/images/map_biomass_total_ us.jpg.

Notes: This NREL study estimates the biomass resources currently available in the United States by county. It includes the following feedstock categories: crop residues (five year average: 2003-2007), forest and primary mill residues (2007), secondary mill and urban wood waste (2002), methane emissions from landfills (2008), domestic wastewater treatment (2007), and animal manure (2002).

For more information on the data development, please refer to http://www.nrel.gov/docs/fy06osti/39181.pdf. Although the document contains the methodology for the development of an older assessment, the information is applicable to this assessment as well; the difference is only in the data's time period.

Figure 11. U.S. Biomass Resource Availability.

Technology and Cost Considerations

Combustion technologies used to convert biomass to electricity are generally considered commercial, and there are approximately 80 operating biomass electricity generation facilities located in the United States.[79] Nevertheless, using biomass as a feedstock for electricity generation can be challenging because each biomass type has different properties, such as water content, ash content, and energy value.[80]

This variability in feedstock quality and characteristics typically must be addressed in order to effectively operate biomass electricity generation equipment. Further, some biomass materials contain certain alkali metal species, such as sodium and potassium, that can potentially impede the operation of electricity generation equipment.[81]

EIA estimates that the levelized cost of energy for biomass electricity ranges from $99.50 per MWh to $133.40 per MWh. Biomass accumulation and transportation and biomass feedstock quality might be considered key cost drivers that can impact the levelized cost of energy for biomass electricity.[82]

Table 5. Annual U.S. Biomass Electricity Generation Potential

% of 1.3B	10%			50%		100%
tons	Low	High	Low	High	Low	High
Electricity Generation (GWh)	125,730	142,880	628,660	714,390	1,257,330	1,428,780

Source: CRS analysis of scenarios based on, "Biomass as Feedstock for a Bioenergy and Bioproducts Industry: The Technical Feasibility of a Billion-Ton Annual Supply," U.S. Department of Energy and U.S. Department of Agriculture, April 2005, available at http://www1.eere.energy.gov/ biomass/pdfs/ final_billionton_vision_report2.pdf.

Notes: Calculations for this table were based on the following: (1) annual tons consumed for electricity generation, (2) energy content (Btu) per ton of biomass, and (3) biomass-to-electricity conversion efficiency. Estimates of annual tonnage were based on the percentages listed in the table (10%, 50%, and 100%). Energycontent per ton of biomass was assumed to be 15 million Btu/ton. Biomass-to-electricity conversion efficiency ranged from 22% to 25%. This range is the reason for "low" and "high" estimates in the table. This conversion efficiency is generally representative of biomass combustion technologies, which have a commercial operating history. Other conversion technologies, such as certain gasification or biological conversion approaches, may have different conversion efficiencies. Since there are competing uses for biomass material, it is unlikely that 100% of the potential biomass resource will be used for electricity generation. The "100%" scenario presented in this table is provided for reference only.

Figure 12 provides a comparison of costs for conventional (fossil and nuclear) and renewable electricity generation.

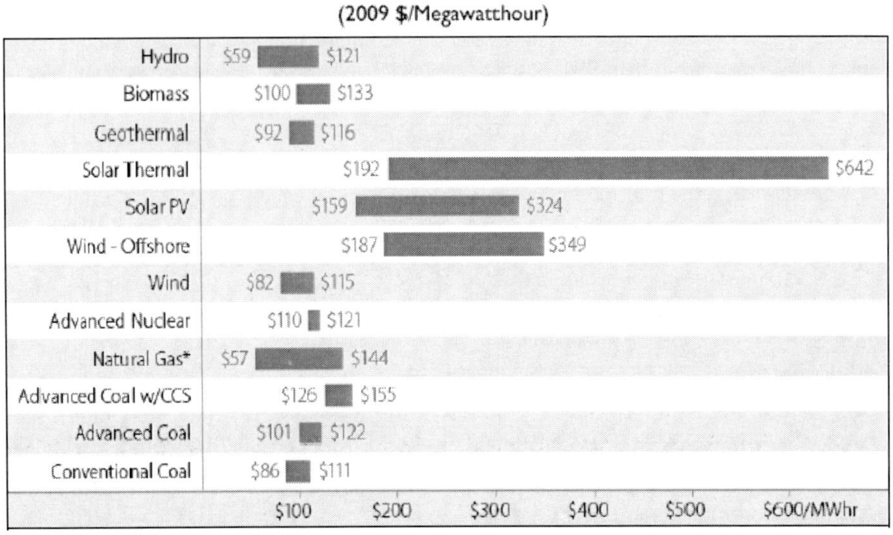

Source: CRS adaptation of EIA's "Levelized Cost of New Generation Resources in the Annual Energy Outlook 2011," available at http://www.eia.gov/oiaf/aeo/ electricity_generation.html.

Notes: EIA LCOE estimates are for new projects that are would be brought on line in 2016. LCOE estimates do not incorporate any federal or state tax incentives.

* The LCOE range for Natural Gas includes four different technologies: (1) conventional combined cycle, (2) advanced combined cycle, (3) conventional combustion turbine, and (4) advanced combustion turbine.

Figure 12. EIA's Levelized Cost of Energy (LCOE) Estimates for New Plants.

Challenges for Renewable Energy

Each type of renewable energy technology has certain advantages and disadvantages relative to each other and relative to fossil fuel energy sources. An extensive literature exists on these advantages and disadvantages.[83] This part of the report offers brief observations and discussion of certain challenges that might affect the full development and deployment of renewable electricity generation technologies. Many policies directed at renewable energy deployment are designed to address these challenges.

Cost

Perhaps the most fundamental challenge to the deployment of renewable energy is the cost of generating electricity from renewable sources. Energy producers and consumers seek the lowest-cost energy, and fossil fuels have historically been the lowest-cost sources of energy, either through end-use combustion or through the generation of electricity.

Levelized Cost of Energy (LCOE)

A common metric for measuring the financial cost of electricity production is Levelized Cost of Energy, or LCOE.[84] LCOE calculations are typically expressed in terms of dollars per unit of energy. The most common units of energy used for comparing the LCOE of different energy sources are kilowatthour (kWh) and megawatthour (MWh). LCOE estimates can provide a relative comparison of energy generation costs for different energy sources such as coal, natural gas, wind, solar, and others. However, policy makers may want to exercise caution when reviewing and considering LCOE estimates. Reasons for this caution include the following: (1) no agreed-upon or standardized LCOE calculation methodology exists, and methods can be tailored to skew results in favor of a particular technology or resource, (2) assumptions used to calculate LCOE estimates can have a major impact on calculations results, and (3) LCOE estimates may not reflect the variable time-of-day value of electricity generation. For example, electricity at 2 p.m. may have more value than electricity at 2 a.m. Therefore, it is unlikely that LCOE calculations performed by different organizations will be identical. Understanding the methodology and assumptions used is often critical when considering LCOE estimates.

Although there is no standard LCOE calculation method, two fundamental methods are commonly used. One method uses total life cycle costs (capital, operations and maintenance, etc.) and total life cycle energy production to calculate a $/kWh or $/MWh cost of energy.[85] Another method uses a project cash flow model to calculate equity rates of return based on the price of energy paid to the project. Typically, a target equity rate of return, expressed as a percentage, is established and the price per unit of energy is adjusted in order to reach the equity return target.[86] This cash-flow-based methodology is unique in that it may include specific project finance constraints such as debt service coverage ratios, cash reserves, and other factors that may not be reflected in the cost vs. energy production approach.

Furthermore, differences in several key assumptions can significantly alter calculations of LCOE estimates. Assumptions that can impact LCOE estimates include (1) capital costs, (2) operation and maintenance costs, (3) government incentives, (4) capacity factor, (5) financial structure (debt/equity ratio), (6) financial costs for debt and equity, (7) project lifetime, and (8) technology performance degradation. Several key assumptions must be included in each calculation of LCOE estimates. Since different organizations often use different assumptions, the variation in LCOE estimates is not surprising.[87]

For this report, LCOE estimates from the EIA Annual Energy Outlook (AEO) 2011 were used to compare the cost of new electricity generation for various renewable energy resources. *Figure 12* summarizes EIA's range of LCOE estimates for several technologies.[88]

It is important to note that EIA LCOE estimates reflect only the projected amount of capacity expected to be added to the electricity generation system during the forecast period. Costs for renewable electricity typically follow a supply curve where costs increase as new capacity is installed, which indicates that the lowest-cost capacity will be added first. Geothermal supply curve estimates provide an example to consider. Figure 13 shows a supply curve for enhanced geothermal electricity costs, developed by NREL. As indicated in the figure, depending on the amount of geothermal capacity installed, the projected LCOE could be as high as $1,000 per MWh. This example of how energy costs can change, as capacity additions increase, further illustrates the importance of understanding all assumptions used for projecting future electricity costs.

Comparing Fossil and Renewable Energy Costs

The current comparatively low cost of fossil fuel energy, as indicated in EIA's LCOE estimates, may not include any costs associated with the external impacts of fossil fuel consumption, which has been stated this way:

> But some energy costs are not included in consumer utility or gas bills, nor are they paid for by the companies that produce or sell the energy. These include human health problems caused by air pollution from the burning of coal and oil; damage to land from coal mining and to miners from black lung disease; environmental degradation caused by global warming, acid rain, and water pollution; and national security costs, such as protecting foreign sources of oil.[89]

Accurately quantifying social costs of fossil energy associated with health problems, climate change, and others can be difficult and complex. As with all cost calculations, assumptions used for estimating social costs can have a dramatic effect on calculation results. Nevertheless, some groups do attempt to place a value on the social costs of fossil energy as an alternative method for comparing the cost of energy from fossil and renewable resources.[90] Furthermore, an Interagency Working Group was created under Executive Order 12866 to estimate the social cost of carbon for regulatory impact analysis.[91] Over time, the costs of mitigating some of these social costs may be placed on the producers or consumers of fossil fuels.

Technology and cost are closely related because renewable energy developers seek technologies that produce energy as inexpensively as possible in order to attain commercially viability. Today, research continues to identify new, more efficient materials and to seek technologies that can be manufactured at lower cost. Improvements continue as new technologies emerge and evolve.

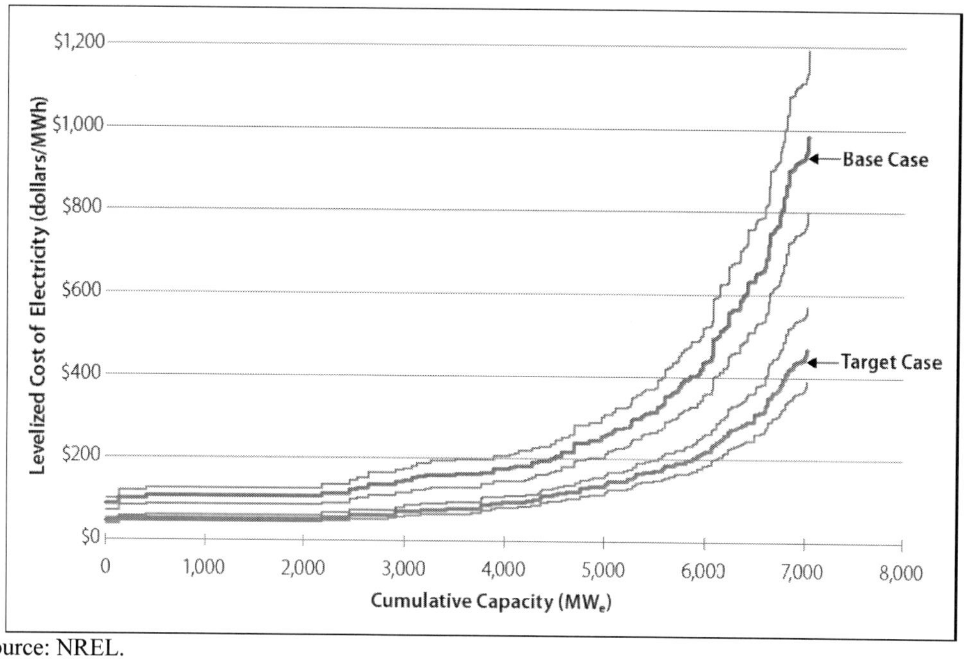

Source: NREL.

Notes: For more information see C. Augustine, K. Young, and A. Anderson, "Updated U.S. Geothermal Supply Curve," National Renewable Energy Laboratory, Conference Paper presented at Stanford Geothermal Workshop, February 1, 2010, available at http://www.nrel.gov/docs/fy07osti /41073.pdf.

Figure 13. NREL Supply Curve for Near-Hydrothermal Field Enhanced Geothermal Systems (EGS) Resource.

Much of the current R&D on renewable technologies aims to reduce the manufacturing cost and the electricity production cost, thereby making renewable electricity more competitive in the marketplace.

Power System Integration

Connecting renewable electricity generation facilities to the electric power grid can raise potential technical challenges. In particular, a high percentage penetration of variable sources—such as solar and wind—can cause serious power quality and reliability problems. The power system requires constant, 24/7 minute-by-minute monitoring and control. The introduction of variable electricity generation may pose power system reliability challenges associated with moment-bymoment balancing of electricity supply and demand.[92] A recent study by the International Energy Agency (IEA) indicates that many existing power systems currently have infrastructure and processes to manage some degree of variability, and these existing assets could potentially be used to manage variable renewable energy resources.[93] However, not all renewable sources of electricity are classified as variable. Biomass, geothermal, and some hydropower sources have the ability to generate electricity on a consistent and predictable basis. As a result, integrating these renewable sources into the power system may not be difficult. However, the inherently variable nature of wind, solar, and some ocean-hydrokinetic electricity may result in significant power system operational challenges if these variable renewable energy sources achieve a high percentage level of penetration.[94]

DOE funded two studies—the Eastern and Western grid interconnection studies—to evaluate the challenges and opportunities associated with significant penetration of variable renewable sources of electricity.[95] The Eastern Wind Integration and Transmission Study assessed the impacts of integrating wind electricity generation at a 20% to 30% penetration level.[96] The Western Wind and Solar Integration Study evaluated potential power system impacts associated with a 35% penetration comprised of wind (30%) and solar (5%).[97] Both studies concluded that integrating the respective penetration rates of variable renewable electricity is manageable, although accommodating those penetration levels may require large amounts of transmission investment, additional reserve capacity, and modifications to power system operations.

Intermittency and Variability

Some renewable energy sources are intermittent and variable. Geothermal, biomass, and some hydropower energy sources usually can be delivered continuously over time. However, wind power is usable only when the wind blows, solar power is usable only when the sun shines, and some hydroelectric power is usable only when water is available to flow through the turbines, so the production of renewable electricity from those sources varies over a period of minutes, hours, days, or months. In addition, wind speed may vary over a period of seconds, minutes, or hours, and solar energy may vary with cloud cover over a period of minutes or hours. This intermittent and variable nature of renewables contrasts with fossil and nuclear power plants, which produce electricity continuously and uniformly except during times of maintenance, fuel supply disruptions, operational problems, or natural disasters. The intermittency and variability of renewable energy might be partially overcome through the development of advanced storage technologies that provide storage of various quantities of electrical energy for use during renewable energy down time. A wide range of batteries, compressed-air storage, hydrogen generation and fuel cells, and other means of storing and recovering intermittent energy are being studied. Such storage is currently costly, and the combination of renewable electricity generation and reliable storage—or backup reserve capacity from natural gas or other dispatchable sources— will need to be considered by the electrical delivery system in order to maximize the potential contributions of renewable technologies.[98]

Renewable Energy Footprint and Land-Use

Although the amount of renewable energy available from the sun, wind, and water may seem unlimited, the land available for energy development is potentially limited by a number of factors. As mentioned above, these sources are dispersed, and technologies are required to convert the natural form of energy into electricity. For this reason, certain renewable energy technologies available today require large areas of land—a large footprint—for each unit of energy produced.

Figure 14 displays different estimates of the land-use intensities of several energy production technologies.

Estimates of the land-use intensity for renewable and nonrenewable sources of energy vary significantly, depending on a number of assumptions. To date, there is no standard methodology to produce these estimates. For example, the extent to which an energy site exclusively "uses" an amount of land is debatable. Only a small portion of area within a wind energy site is actually occupied by the turbines, so remaining land could potentially be—and often is—dedicated to other uses. In contrast, fields of energy crops to be burned in the production of electricity will fully occupy their allotted area. Further, energy production for renewables varies substantially with geography. A solar photovoltaic plant of a certain capacity will require less land if located in a region with more intense sunlight. In addition, it is difficult to compare land-use intensities for renewable energy technologies with those of fossil fuel technologies. For example, for fossil fuels, calculations of land-use intensity may include the power plant footprint, plus mining or production area, plus areas occupied by transportation and logistics infrastructure. Thus, the footprint for natural gas may include the gas power plant, but also the areas occupied by gas wells, the roads that connect the gas wells, and the pipelines that transport the gas to market. Also, the areal extent of infrastructure may not fully represent the impact on the landscape. The degree to which such infrastructure divides or dissects ecosystems may also be an important consideration. [99]

The electric energy production technologies with the greatest land-use intensity (amount of land per unit of electrical energy produced) are biomass, wind, hydropower, and solar photovoltaic.

Land-use intensities of natural gas, coal, geothermal, and nuclear power are likely significantly smaller than those of other forms of energy production. As demand grows for utility-scale installations of renewable energy, pressure will grow to integrate energy policy with land-use policy.[100] The integration of distributed generation technologies, such as rooftop solar, into existing building structures will help mitigate land-use issues, but there will likely remain a strong need for utility-scale renewable energy installations.

Transmission Availability and Access

Though renewable energy technologies may be used across most of the nation, optimized use of renewable energy must accommodate certain geographic controls. The wind energy resource is richest in coastal areas and the Midwest. Solar energy is optimal in the relatively cloudless southwestern United States.

Hydroelectric power has historically been best deployed on large rivers with steep gradients. These geographic concentrations of renewable energy sources often mean that the energy may be optimally produced far from the existing energy demand centers, which are the large cities of the east and west coasts, upper Midwest, and South. Thus, large-scale deployment of renewable energy technologies will likely be accompanied by the need for new electricity transmission infrastructure from the new regions of energy supply to the demand centers.[101] For example, the NREL Eastern Interconnection Report concluded that 20% to 30% wind generation is feasible, but would require "significant expansion of the transmission infrastructure."[102]

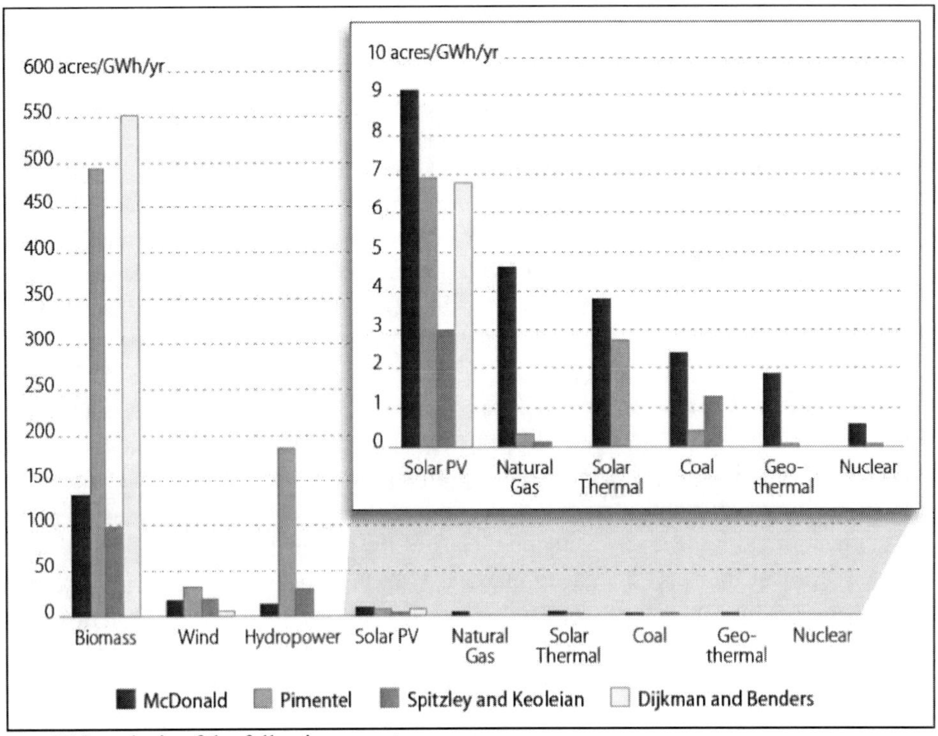

Source: CRS analysis of the following reports:

- McDonald RI, Fargione J, Kiesecker J, Miller WM, Powell J(2009) Energy Sprawl or Energy Efficiency: Climate Policy Impacts on Natural Habitat for the United States of America. PLoS ONE 4(8): e6802. doi:10.1371/journal.pone.0006802, Figure 3.
- David Pimentel et al., "Renewable Energy: Current and Potential Issues," BioScience, vol. 52, no. 12 (December 2002), pp. 1111-1120.
- David V. Spitzley and Gregory A. Keoleian, Life Cycle Environmental and Economic Assessment of Willow Biomass Electricity: A Comparison with Other Renewable and Non-Renewable Sources, Center for Sustainable Systems, Report No. CSS04-05R, Ann Arbor, MI, March 25, 2004 (revised February 10, 2005).
- T.J. Dijkman and R.M.J. Benders, "Comparison of renewable fuels based on their land use using energy densities," Renewable and Sustainable Energy Reviews, vol. 14 (2010), pp. 3148-3155.

Notes: GWh = gigawatthours, yr = year; Acres per gigawatthour per year (GWh/yr) is the metric used to compare results from the respective reports. GWh/yr indicates the amount of land required to generate a certain amount of electricity, in this case a gigawatthour. Some studies report land use per unit of capacity, which might be reported as acres per gigawatt (GW). Land use per capacity is somewhat misleading because each energy technology has a different capacity factor, meaning that operational hours for each technology will vary over the course of a year. Reflecting land use as a function of electricity generation takes into account capacity factor differences.

Figure 14. Land-Use Intensity for Various Forms of Energy Production.

Not only must a new installation of renewable energy technology be connected to the grid, but the new transmission infrastructure must be sized to the maximum rate of electricity flow even though it may flow intermittently at that rate.

Materials and Resources

While renewable energy sources may provide a virtually infinite supply of energy, building and installing the equipment necessary to convert renewable energy into usable electricity may require significant quantities of materials and other natural resources. For example, wind turbine manufacturing requires a number of materials and resources, the most critical being steel, fiberglass, resins, blade core materials, permanent magnets, and copper.[103] Current solar photovoltaic technologies require materials such as silicon, cadmium, tellurium, silver, and others.[104] Large-scale wind and solar deployment would raise demand for these materials, which in turn may impact their respective prices.

This potential price impact may be an important consideration, since the cost of renewable electricity generation is highly correlated with the cost of the energy conversion system (i.e., wind turbines, solar panels, etc.).

Environmental Impact and Aesthetic Concerns

Capturing and converting any energy source—including renewable energy—will have some degree of impact on the environment. Land use and habitat disturbance are potential environmental issues for wind and solar electricity projects. Installation of wind turbines has already attracted attention because of bird mortality, noise, and resulting NIMBY[105] attitudes. Some CSP technologies may require vast amounts of water, although dry-cooling CSP technologies, with lower efficiencies, are available.[106] Water use, land subsidence, and seismicity may need to be addressed by geothermal power plants. Hydropower and ocean-hydrokinetic electricity generation systems may result in water quality degradation, ecosystem disruption, and animal mortality. Biomass projects impact the environment through emissions such as nitrogen oxides (NO_x), carbon dioxide (CO_2), and others, as well as land use changes associated with producing biomass feedstock.[107] These, and other, potential environmental impacts may need to be considered as policy makers look to balance the desire to increase electricity production from renewable sources of energy with environmental objectives.

Infrastructure Requirements

All forms of energy production and delivery require some form of infrastructure. Coal is delivered by an extensive network of railroads, and natural gas is delivered via a large network of pipelines. In addition to new transmission requirements, some renewable energy sources may require investments in specialized infrastructure in order to provide a source of renewable electricity.

One example is offshore wind energy. Specialized vessels, purpose-built portside infrastructure, undersea electricity transmission lines, and grid interconnections will likely be required to support offshore wind development. According to DOE, "these vessels and this infrastructure do not currently exist in the U.S."[108] Such specialized infrastructure requirements may also be a consideration for policy decisions associated with certain other types of renewable energy.

Technology Development and Commercialization

Some renewable electricity generation technologies are not yet commercially available. Private and public investments are being made in renewable electricity generation technologies, to include venture capital firms, private and public corporations, and the U.S. DOE through its Advanced Research Projects Agency—Energy (ARPA-E) program office. While ARPA-E and DOE's Office of Energy Efficiency and Renewable Energy provide funds to support technology R&D, concept demonstrations, and technology performance optimization, bridging the gap between these activities and commercialization may require significant amounts of funding. Commonly known as the commercialization "valley of death," several additional market development activities that might include technology performance characterization and validation, operational reliability assessments, accurate quantification of maintenance and operations costs, etc., may be necessary in order for new technologies to qualify for private equity and bank/debt finance in support of commercial projects. Obtaining the funds necessary to commercialize new technologies can be difficult and costly.[109]

Policy and Regulatory Challenges

Certain federal and state-level policies have served to stimulate growth of renewable electricity generation. Federal policies such as production and investment tax credits for certain renewable energy property, along with other tax-favored finance options, have created financial incentives for building and operating renewable electricity generation projects.[110] Other federal financial incentives are also available for renewable energy.[111] Furthermore, the American Recovery and Reinvestment Act (ARRA) provided new policies such as the Section 1603 cash grant option for renewable electricity generation projects and the Section 1705 Loan Guarantee Program that provides government-backed debt financing for certain renewable energy projects.[112]

State-level policies, such as renewable portfolio standards, have served to create market demand for renewable electricity.[113] Furthermore, there is some interest in establishing a federal renewable or clean energy standard, which may create additional demand for renewable electricity generation.[114]

Federal policies that support renewable electricity generation typically are available for a defined period of time, at the end of which the policies expire. Some in the renewable electricity industry argue that the sudden expiration of certain federal policies has resulted in market uncertainty and downward pressure on renewable electricity market growth.[115]

The historical start-stop nature of federal policies may be challenging to the renewable energy industry due to a lack of long-term financial certainty for renewable electricity generation projects. On the other hand, some policy makers may not wish to create a policy environment that results in a renewable energy industry that is dependent on federal financial incentives. Balancing policy objectives that might stimulate a solid base for renewable electricity, while at the same time eliminating a dependency on federal subsidies, may be a consideration for policy makers.

Related Issues

Energy Efficiency and Curtailment

Although this report does not provide a detailed analysis of energy efficiency and conservation, it is widely acknowledged that both energy efficiency (doing as much or more with less energy and eliminating waste) and curtailment of demand (doing less with less energy) provide enormous opportunities for reducing or controlling the energy resources of the nation. By addressing the demand side of the energy equation, as well as the supply side, the United States can extend the energy resources that it consumes. Efficiency and demand curtailment will not, by themselves, meet the demand for energy in the future, but these strategies will likely reduce the amount of new energy needed.[116]

The benefits of more efficient use of energy are being sought by a wide range of citizens, homeowners, manufacturers, and governments. Lower costs, reduced greenhouse gas emissions, and reduced need for expansion of supply are key motivators to increase energy efficiency and conservation. Energy efficiency can be measured for individual devices such as appliances, automobiles, and light bulbs, but derivative indicators are used to measure levels and trends in energy efficiency at a national level. The most common national indicator of energy efficiency and curtailment is energy intensity.[117] Energy intensity is measured in units of energy per dollar of Gross Domestic Product (GDP). As *Figure 14* shows, the energy intensity of the United States has been dropping steadily for decades, despite the steady growth in total energy consumption. There are many reasons for this trend, of course, including a gradual change from a manufacturing economy to a more service-oriented economy, but ongoing efforts to promote energy efficiency and conservation are clearly succeeding in the United States. For example, one study estimates that improving the energy efficiency of buildings in the United States could save $170 billion per year in energy costs through 2030.[118] Numerous other opportunities exist for improving efficiency or curtailment in energy use in the United States.[119]

Biofuels

Biofuels are liquid fuels produced from plant materials, which makes them a renewable commodity. The major biofuels are fuel ethanol and biodiesel, though other kinds of alcohols and hydrocarbons can also be synthesized from biological materials. Both fuel ethanol and biodiesel are currently used primarily as blending agents with conventional gasoline and diesel fuel, though both can conceivably be used in their pure form with some modifications to engine fuel systems.[120]

Unlike the other kinds of biomass discussed above, liquid biofuels are normally used as transportation fuels and are not used to generate electricity. Liquid biofuels are important because certain forms of transportation such as aircraft and heavy trucks cannot easily be converted to electricity or other propulsion technologies. In 2009, the United States consumed 99 million gallons of fuel ethanol as an 85% blend (E85), 10.7 billion gallons of fuel ethanol as a 15% blend (E15) in gasoline, and 316 million gallons of biodiesel.[121]

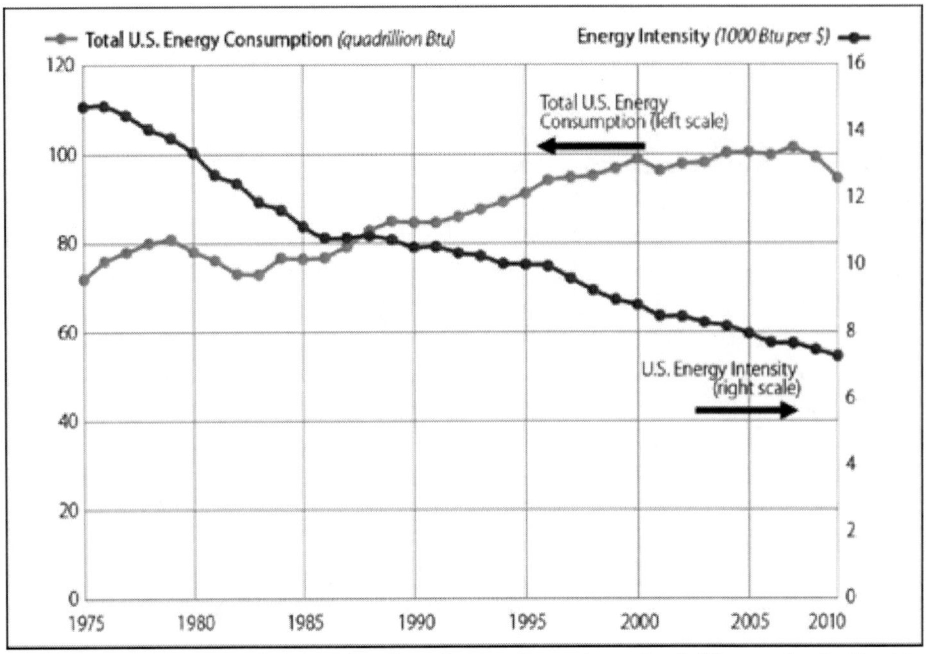

Source: Energy Information Administration, http://www.eia.gov/emeu/aer/pdf/ pages/sec1_13.pdf.

Note: Energy intensity is the total primary energy consumption per real dollar of Gross Domestic Product.

Figure 15. Total U.S. Energy Consumption and Energy Intensity, 1975-2009.

Additional Considerations for Renewable Electricity in the United States

The Scale of U.S. Energy Consumption

One important aspect of the expansion of renewable forms of energy, often overlooked or under-appreciated, is the scale or magnitude of energy use in the United States. It is not only *what kind* of energy is used, but *how much* energy the United States uses on a daily, monthly, and annual basis. By any measure, the amounts of energy used by the United States are prodigious, and replacing a significant proportion of fossil fuels with renewable forms of energy would be a formidable task. Alternatives to fossil fuels must be produced on a very large scale and must be available to all parts of the nation to provide the enormous and increasing amounts of energy demanded by the U.S. economy. Whether individual renewable energy installations are large (utility-scale) or small (distributed), the total combined output must accommodate the very large—and increasing—demand for energy. Current electricity generation is dominated by coal, natural gas, and nuclear (see Table 6). Only 8% of total energy use in the United States is renewable, and 53% of that is for electricity generation. In 2009, total U.S. energy use was 94.6 quadrillion Btu, and renewable electricity accounted for about 4 quadrillion Btu.[122] Therefore, any serious proposal to displace fossil fuels with renewable energy must include massive growth in renewable energy technology deployment.

Table 6. Total U.S. Electricity Generation, By Source, 2009

Generation fuel	GWh	%
Coal	1,755,904	44.45
Petroleum	38,937	0.99
Natural Gas	920,979	23.31
Other Gases	10,632	0.27
Nuclear	798,855	20.22
Hydroelectric Conventional	273,445	6.92
Wind	73,886	1.87
Solar Thermal and Photovoltaic	891	0.02
Wood and Wood Derived Fuels	36,050	0.91
Geothermal	15,009	0.38
Other Biomass	18,443	0.47
Pumped Storage	-4,627	-0.12
Other	11,928	0.30
All Energy Sources	3,950,332	100.00

Source: EIA, http://www.eia.doe.gov/cneaf/electricity

Notes: Electricity generation from pumped storage is negative since pumping water into a storage reservoir requires more electricity than that generated when the stored water is used to operate a turbine. Pumped storage projects are typically based on opportunities to pump water into a reservoir when electricity prices are low (typically at night), then use the stored water to generate electricity when prices are high (typically during peak demand hours).

Relationship between Renewable Electricity and Imported Energy

Petroleum consumption may be displaced by the production of biofuels, but most renewable energy technologies are designed to generate electricity. Therefore, the use of renewable energy to generate electricity in today's U.S. market would displace only those fossil fuels that are used to generate electricity, and the United States uses almost no imported fossil fuels to generate electricity. For example, the U.S. transportation system is 94% reliant on petroleum (*Figure 1*), and the use of renewable electricity for transportation might require increased electrification of the transportation system. Consequently, the only way that increasing production of renewable electricity would affect oil imports is if the U.S. transportation system is electrified so that domestically generated electricity substitutes for oil. Likewise, any process in which the burning of natural gas is used for direct heating would need to be electrified in order for renewable energy to substitute. More than 93% of U.S. coal consumption is used to generate electricity, so adopting renewable energy sources to generate electricity could potentially reduce demand for coal, but would have no effect on energy imports because virtually all of U.S. coal is produced domestically.

International Renewable Electricity Markets

Recent news reports emphasize how successful China and other nations have become in developing and deploying renewable energy technologies. Indeed, China is constructing impressive amounts of renewable energy installations, but the United States remains one of the world leaders in renewable energy capacity and deployment. For example, *Table 7* shows that the United States leads the world in installed non-hydropower renewable electricity generation capacity, biomass power, and geothermal power. While the United States ranked second, behind China, in total wind power capacity, the United States ranked first in 2010 in terms of *operational* wind power capacity.[123] In 2009, the United States generated more electricity from non-hydro renewable energy sources than any other country in the world. However, when hydropower is included, China led the world in terms of total renewable electricity generation (see Figure 16).

Table 7. Existing Renewable Energy Capacities at the End of 2010 (Country ranking for selected categories)

Rank	Renewablespower capacity (not including hydro)	Renewables power capacity (including hydro)	Wind power	Biomass power	Geothermal power	Solar PV	Solar hot water/heat
1	United States	China	China	United States	United States	Germany	China
2	China	United States	United States	Brazil	Philippines	Spain	Turkey
3	Germany	Canada	Germany	Germany	Indonesia	Japan	Germany
4	Spain	Brazil	Spain	China	Mexico	Italy	Japan
5	India	Germany/ India	India	Sweden	Italy	United States	Greece

Source: REN21. 2011. Renewables 2011 Global Status Report (Paris: REN21 Secretariat), available at http://www.ren21.net/REN21Activities/Publications /GlobalStatusReport/GSR2011/ tabid/56142 /Default.aspx.

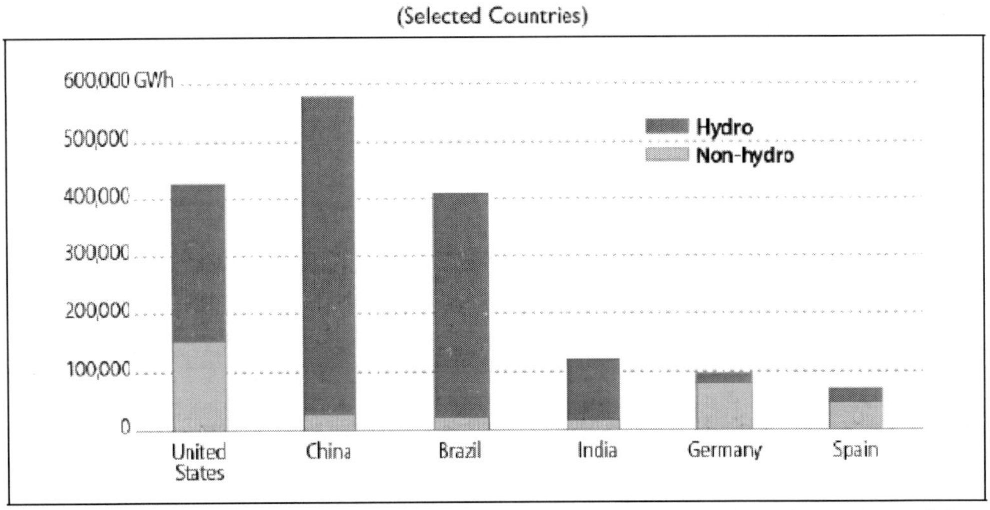

Source: Energy Information Administration, International Energy Statistics, http://www.eia.gov/cfapps/ ipdbproject/IEDIndex3.cfm?tid=6&pid=29&aid=12.

Notes: Non-hydro includes generation from wind, solar, geothermal, tide and wave, and biomass and waste.

Figure 16. Total Net Renewable Electricity Generation, 2009.

FUTURE TRENDS IN RENEWABLE ELECTRICITY

The potential for renewable electricity generation in the United States is very large, yet current use of renewable energy for electricity production is relatively modest, constituting only 11% of total electricity generation and 8% of total energy consumption. Based on the current status of renewables in the United States, policy makers may consider some key questions about the future of renewable energy:

- Should the United States actively seek greater use of renewable energy to supply electricity, or should the energy and electricity markets be allowed to work without further interference with the existing structure of subsidies and incentives?
- If greater use of renewable energy for electricity is desired, what are the key barriers or actions that should be addressed by federal policy?

Future trends in renewable electricity will depend heavily on the cost of both renewable technologies and fossil fuel costs, and on government incentives for renewable energy. In the absence of subsidies for renewable electricity technologies, and in the absence of accounting for external costs of using fossil fuel combustion to generate electricity, several renewable electricity technologies are currently not commercially viable, or only marginally so. Reference case projections by EIA of growth in wind and solar electricity to 2035 are predicated on the use of renewable portfolio standards, renewable fuel standards, and subsidies in the tax code.[124] With low coal and natural gas prices, and high renewable energy technology costs, and the absence of regulation or subsidies, renewable electricity may not increase significantly. Without some form of carbon pricing or other consideration of the externalities of fossil fuel combustion, the United States may remain in an era of relatively low-cost fossil fuel electricity for decades.

However, policy makers may decide that growth in renewable electricity is desirable because of concerns about greenhouse gas emissions and climate change, because fossil fuel supplies are ultimately finite, and because of a desire to position the United States as a global leader for renewable energy technology and manufacturing. Renewables could be made more cost competitive by means of improved renewable technologies or revised cost of carbon-based fuels, but financial or regulatory incentives may be required to make certain renewable sources more economically viable in the short term.

In the event that levelized costs of renewable electricity become competitive with those of fossil fuel electricity, the additional issues of intermittency/variability, land-use and footprint, the need for additional transmission, plus other resource and environmental impacts of renewable electricity will need to be addressed by local, state, and federal officials and policy makers.

CONCLUSION

Cumulative U.S. renewable electricity generation capacity more than doubled from 2006 to 2010, increasing from approximately 22 GW to nearly 55 GW.[125] In 2010, renewable sources of energy provided approximately 11% (7% from hydropower and 4% from other renewables) of total net electricity generation and the EIA AEO 2011 reference case projects that renewable electricity generation will increase to between 14% and 15% by 2035.126 The renewable electricity generation research conducted for this report indicates that the potential may exist for renewable energy sources to make a sizeable contribution toward total U.S. electricity generation demand. However, renewable electricity generation will likely encounter serious challenges, issues, and barriers as technologies and projects look to realize large-scale deployment. As Congress evaluates various energy policy objectives, policy makers may move to holistically evaluate the potential intended benefits, such as emissions reduction and job creation, with potential risks and consequences, such as electricity cost/price increases and electricity delivery reliability issues associated with increasing renewable electricity generation.

End Notes

[1] Energy Information Administration, Annual Energy Review 2009, http://www.eia.doe.gov/ totalenergy/ data /annual/pdf/pecss_diagram_2009.pdf.

[2] Decreasing U.S. reliance on foreign oil is not included here because the focus of this report is on electricity generation. Petroleum contributed 1% of electricity generation in 2009. Based on current U.S. energy infrastructure, adding additional renewable electricity generation capacity will have a negligible, if any, impact on U.S. oil import dependency. However, electrification of the transportation fleet could potentially result in decreasing total U.S. oil demand. Renewable electricity generation combined with electric vehicle market penetration could potentially result in lower oil import requirements.

[3] Biomass and biofuels release CO_2 during combustion, but are considered by some to have zero net emissions because the CO_2 released was taken up from the atmosphere to grow the plants. However, there is debate about biomass being considered carbon neutral. For more information see CRS Report R41603, *Is Biopower Carbon Neutral?*, by Kelsi Bracmort.

[4] CRS is aware that the National Renewable Energy Laboratory (NREL) is in the process of publishing an analysis about U.S. renewable electricity generation potential. However, the NREL work was not available to influence the research for this report.

[5] Energy Information Administration, http://www.eia.gov/energyexplained/index.cfm? page= renewable_home.

[6] Energy Information Administration, http://www.eia.gov/state/state-energy-profiles-analysis. cfm?sid=WA.

[7] A British thermal unit (Btu) is a measure of the energy (heat) content of fuels. It is the quantity of energy (heat) required to raise the temperature of 1 pound of liquid water by 1°F at the temperature that water has its greatest density (approximately 39°F), http://www.eia. doe.gov/energyexplained/index.cfm?page=about_btu.

[8] A kilowatthour is equal to one thousand watthours. 1 kWh = 3,412 Btu.

[9] A gigawatthour is equal to one billion watthours.

[10] For more information on U.S. fossil fuel resources, see CRS Report R40872, *U.S. Fossil Fuel Resources: Terminology, Reporting, and Summary*, by Gene Whitney, Carl E. Behrens, and Carol Glover.

[11] "Assessment of Moderate- and High-Temperature Geothermal Resources of the United States," U.S. Geological Survey, 2008, available at http://pubs.usgs.gov/fs/2008 /3082/pdf/ fs2008-3082.pdf, and "The Future of Geothermal Energy: Impact of Enhanced Geothermal Systems (EGS) on the United States in the 21st Century," Massachusetts Institute of Technology, 2006, available at http://geothermal.inel.gov/publications/ future_of_ geothermal_energy.pdf.

[12] Resource estimate changes are not unique to renewable energy. Fossil fuel estimates typically change in response to technology, economic conditions, improved data sets, etc. Shale gas in the United States is an example of how resource estimates can change over time.

[13] For more information about the Annual Energy Outlook reference case see, Energy Information Administration, "Annual Energy Outlook 2011," Report Number: DOE/EIA-0383(2011), April 2011, available at http://www.eia.gov/forecasts/aeo/pdf/0383(2011).pdf.

[14] In February 2010, NREL and AWS Truepower released estimates for windy land area and wind energy potential for the 48 contiguous United States. A revision to these estimates that includes data for Alaska and Hawaii was released in April 2011. This is the first comprehensive update of wind energy potential since 1993. The NREL

AWS study evaluates three gross (no system losses included) capacity factor assumptions (30%, 35%, and 40%) and two hub heights (80 meters and 100 meters). NREL/AWS also considered certain land area exclusions such as parks, urban areas, and others. Study results, maps, and data tables available at http://www.windpoweringamerica.gov/wind_maps.asp

[15] NREL's offshore wind study does not take into account any potential area exclusions. The offshore wind study also does not include estimates for Florida, Mississippi, Alabama, or Alaska. For more information see Marc Schwartz, Donna Heimiller, Steve Haymes, and Walt Musial, "Assessment of Offshore Wind Energy Resources for the United States," National Renewable Energy Laboratory, June 2010, available at http://www.windpower ingamerica.gov/pdfs/offshore/offshore_wind_resource_assessment.pdf.

[16] Gigawatthour estimates for offshore wind energy resources were calculated by CRS by applying average capacity factor assumptions of 30% and 40% to megawatt installed capacity estimates from NREL.

[17] "A National Offshore Wind Strategy: Creating an Offshore Wind Energy Industry in the United States," U.S. Department of Energy, Energy Efficiency and Renewable Energy, February 2011, available at http://www1.eere.energy.gov/windandhydro /pdfs /national_ offshore_wind_strategy.pdf.

[18] For the purpose of this report, "commercially available" refers to renewable electricity generation technologies that have achieved an adequate amount of operational time that allows for performance validation, accurate reliability assessments, and an understanding of actual operations and maintenance requirements. These commercialization parameters are typically validated by an independent engineering firm. This independent validation is typically necessary for technologies, and projects that use these technologies, to obtain debt and equity for project development. Furthermore, commercially available technologies typically have an established supply chain of companies that can provide equipment to meet certain project and technology performance specifications.

[19] American Wind Energy Association (AWEA) U.S. Wind Industry Annual Market Report Year Ending 2010.

[20] More information on DOE's Wind Power program is available at http://www1.eere.energy.gov/windandhydro /wind_power.html.

[21] Levelized Cost of New Generation Resources in the Annual Energy Outlook 2011, U.S. Energy Information Administration, available at http://www.eia.gov/oiaf/ aeo/electricity _generation.html.

[22] An overview of solar resources is provided by the Department of Energy at http://www.eere.energy.gov/ basics/renewable_energy/solar_resources.html.

[23] For a comprehensive summary of current solar resource assessment information, see "Report to Congress on Renewable Energy Resource Assessment Information for the United States," U.S. Department of Energy, Office of Energy Efficiency and Renewable Energy, 2011.

[24] The approach used to quantify annual solar electricity generation potential was based on input from experts at NREL and Sandia National Laboratory. If different methodologies for calculating solar resource potential, such as quantifying the total amount of solar radiation exposure on the surface area of the United States, are employed, results may be different (likely much higher) than the summary data presented in this section.

[25] For an overview of CSP technology see http://www.eere.energy.gov/basics/renewable_ energy/csp.html. For an overview of PV technology see http://www.eere.energy. gov/basics/renewable_energy/photovoltaics.html.

[26] Three solar radiation components are typically measured and reported: (1) direct beam solar radiation, (2) diffuse solar radiation, and (3) global solar radiation (the sum of direct beam and diffuse). CSP systems are able to use only direct beam solar radiation. PV systems are able to use both direct beam and diffuse radiation. More detail regarding solar radiation components is available at http://www.eere.energy.gov/basics/ renewable_energy/solar_ esources.html.

[27] Solar energy system configurations may include (1) south-facing flat-plate collectors at various tilt angles, (2) oneaxis flat-plate tracking, (3) two-axis flat-plate tracking collectors, and (4) direct-beam one and two-axis tracking concentrating collectors. For more information regarding solar system configurations see, D. Renne, R. George, S. Wilcox, T. Stoffel, D. Myers, and D. Heimiller, "Solar Resource Assessment," National Renewable Energy Laboratory, February 2008, available at http://www.nrel.gov/docs/fy08osti/ 42301.pdf.

[28] For a description of various CSP technologies, see http://www.solarpaces.org/CSP_ Technology/csp_technology. htm. For a description of various PV technologies, see http://solarbuzz.com/ going-solar/understanding/ technologies.

[29] Tom Mancini, "CSP Overview," Sandia National Laboratories.

[30] *Ibid.* Filters applied to derive these analysis results include (1) sites with >6.75 kwh/m2/day direct normal insolation, (2) excluding environmentally sensitive lands, major urban areas, etc., (3) removing land with slope >1%, 4) only including contiguous areas >10km2. Changing these filters (i.e. reducing the direct normal insolation threshold) would yield different results.

[31] *Ibid.* Arizona, California, Colorado, Nevada, New Mexico, Texas, and Utah.

[32] P. Denholm and R. Margolis, "Land-use requirements and the per-capita solar footprint for photovoltaic generation in the United States," Energy Policy 36, 3531-3543, 2008.

[33] For more information see P. Denholm and R. Margolis, "Impacts of Array Configuration on Land-Use Requirements for Large-Scale Photovoltaic Deployment in the United States," NREL, Conference paper presented at SOLAR 2008—American Solar Energy Society (ASES), May 3-8, 2008.

[34] Telephone interview with Robert Margolis at NREL.

[35] P. Denholm and R. Margolis, "Supply Curves for Rooftop Solar PV-Generated Electricity for the United States," National Renewable Energy Laboratory, November 2008, available at http://www.nrel.gov/docs/fy09osti/44073.pdf.

[36] The remaining 2% of installed capacity consists of power tower, linear fresnel, and dish-stirling technologies. See Tom Mancini, "CSP Overview," Sandia National Laboratories.

[37] For more information about crystalline solar cells, see Y.S. Tsuo, T.H. Wang, and T.F. Ciszek, "Crystalline-Silicon Solar Cells for the 21st Century," NREL, May 1999, available at http://www.nrel.gov/docs/fy99osti/26513.pdf.

[38] More information about DOE CSP R&D programs and projects is available at http://www1.eere.energy.gov/solar/csp_program.html. More information about DOE PV R&D programs and projects is available at http://www1.eere.energy.gov/solar/ photovoltaics_program.html.

[39] For more information on electrical energy storage, see "Energy Storage: Program Planning Document," Department of Energy, Office of Electricity Delivery and Energy Reliability, February 2011, available at http://www.oe.energy.gov/DocumentsandMedia/OE_Energy_ Storage_Program_Plan_Feburary_2011v3.pdf. For more information on "smart-grid," see "The Smart Grid: An Introduction," Department of Energy, available at http://www.oe.energy.gov/DocumentsandMedia/ DOE_SG_ Book_Single_Pages(1).pdf.

[40] For more information on EIA assumptions and calculation methodology see http://www. eia.gov/oiaf/aeo/electricity_generation.html.

[41] More information about DOE's SunShot initiative can be found at http://www1.eere. energy.gov/solar/sunshot/.

[45] Phone interview with Chad Augustine at NREL.

[46] Geothermal Energy Association, more information available at http://geo-energy.org/Basics.aspx.

[47] Geothermal Energy Association, more information available at http://geo-energy.org /plants.aspx.

[48] An overview of Enhanced Geothermal Systems (EGS) technology is available at http://www1.eere.energy.gov/geothermal/pdfs/egs_basics.pdf.

[49] The Department of Energy has established EGS commercialization programs. More information available at http://www1.eere.energy.gov/ geothermal/ enhanced_ geothermal_systems.html.

[50] For more information on EIA assumptions and calculation methodology see http://www.eia.gov/oiaf/aeo/electricity_generation.html.

[51] C. Augustine, K. Young, and A. Anderson, "Updated U.S. Geothermal Supply Curve," National Renewable Energy Laboratory, Conference Paper presented at Stanford Geothermal Workshop, February 1, 2010, available at http://www.nrel.gov/ docs/fy07osti/ 41073.pdf. NREL LCOE estimates are in 2008 US$.

[52] Ibid.

[53] Energy Information Administration, more information available at http://www.eia.gov/cneaf/solar.renewables/page/hydroelec/hydroelec.html.

[54] "Conventional" hydropower resource assessments typically include large hydropower dams, increasing capacity at existing facilities, non-powered dams, small hydro, and low-power hydro. Pumped storage hydroelectricity generation potential is not included in the resource estimates included in this report.

[55] Pumped storage generation potential was not included in the resource assessment literature reviewed for this report. However, pumped-storage projects are being developed and the Federal Energy Regulatory Commission (FERC) has issued pre-permits for about 33 gigawatts of pumped storage capacity (see http://www.ferc.gov/industries/hydropower/gen-info/licensing.asp). The business case for pumped storage might be viewed as an arbitrage opportunity whereby water is pumped to a reservoir when energy prices are low, and the stored water is used to generate electricity when energy prices are high or an opportunity exists to receive a financial premium for stand-by or firm power. According to EIA, more energy is required to pump water into a storage reservoir than is generated when electricity is produced by releasing the stored water. However, pumped storage facilities can provide valuable ancillary on-demand energy production services for electricity grid operators. For more information see http://www.ferc.gov/industries/hydropower/ gen-info/regulation/ pump.asp.

[56] A. Conner, J. Francfort, and B. Rinehart, "U.S. Hydropower Resource Assessment Final Report," Idaho National Engineering and Environmental Laboratory, December 1998, available at http://hydropower.inl.gov/resourceassessment/pdfs/doeid-10430.pdf.

[57] D. Hall, R. Hunt, K. Reeves, and G. Carroll, "Estimation of Economic Parameters of U.S. Hydropower Resources," Idaho National Engineering and Environmental Laboratory, June 2003, available at http://hydropower.inl.gov/resourceassessment/pdfs/project_report-final_with_disclaimer-3jul03.pdf.

[58] D. Hall, K. Reeves, J. Brizzee, R. Lee, G. Carroll, and G. Sommers, "Feasibility Assessment of the Water Energy Resources of the United States for New Low Power and Small Hydro Classes of Hydroelectric Plants," Idaho National Laboratory, January 2006, available at http://hydropower.inl.gov/resourceassessment/pdfs/main_report_appendix_a_final.pdf. Note: This report quantified hydropower resource potential as megawatts-annual (MWa) based on a 50% capacity factor assumption. As a result, CRS had to convert MWa estimates to megawatts (MW) in order to have resource estimates on an equivalent basis.

[59] The primary difference between the reports was the inclusion of low-power (<1MW) hydropower resources in the INL 2006 report.

[60] Electricity generation potential assumes a 50% capacity factor.

[61] "Assessment of Waterpower Potential and Development Needs," Electric Power Research Institute, 2007, available at http://www.aaas.org/spp/cstc/docs/07_06_1ERPI_report.pdf.

[62] Presentation by Brennan T. Smith to the National Hydropower Association Annual Conference, "U.S. Hydropower Fleet and Resource Assessments," Oak Ridge National Laboratory, April 5, 2011, available at http://hydro.org/wpcontent/ uploads/2011/04/ Brennan-Smith-PPT_NHA_April2011_Final.pdf.

[63] For more information on the history of hydroelectricity in the United States see http://www1.eere.energy.gov/windandhydro/hydro_history.html.For an overview of hydroelectricity technologies see CRS Report R41089, *Small Hydro and Low-Head Hydro Power Technologies and Prospects*, by Richard J. Campbell.

[64] For more information, see http://www1.eere.energy.gov/windandhydro/printable_ versions/ hydro_advtech.html.

[65] DOE, in April 2011, announced $10.5 million of funding for small hydropower technologies that could be deployed in constructed waterways. For more information see http://www.energy.gov/news/10255.htm.

[66] See EIA "Levelized Cost of New Generation Resources in the Annual Energy Outlook 2011," available at http://www.eia.gov/oiaf/aeo/electricity_generation.html.

[67] For more information regarding these energy production approaches, see "Ocean Energy Technology Overview," Department of Energy, Office of Energy Efficiency and Renewable Energy, July 2009, available at http://www1.eere.energy.gov/femp/pdfs/44200.pdf. Osmotic, or salinity gradient, power is another possible source of ocean energy. However, this energy production source has not yet been explored or analyzed in great detail. Background on osmotic power is available at http://en.wikipedia.org/wiki/Osmotic_power.

[68] For more information about EPRI's wave energy resource assessment, see http://oceanenergy.epri.com/waveenergy.html. For more information about EPRI's tidal energy resource assessment see http://oceanenergy.epri.com/streamenergy.html.

[69] On July 6, 2011 DOE released a database, developed in partnership with the Georgia Institute of Technology, of tidal energy resources in the United States. The interactive database is available online at http://www.tidalstreampower.gatech.edu/.

[70] G. Miller, J. Franceschi, W. Lese, and J. Rico, "The Allocation of Kinetic Hydro Energy Conversion Systems (KHECS) in USA Drainage Basins: Regional Resource Potential and Power," New York University, Department of Applied Science, August, 1986.

[71] Capacity factor estimates for hydrokinetic devices were based on data reported by Argonne National Laboratory. See http://teeic.anl.gov/er/hydrokinetic/restech/scale/index.cfm.

[72] As of June 9, 2011, the Federal Energy Regulatory Commission (FERC) had issued 70 preliminary permits for tidal, wave, and inland hydrokinetic projects. For more information, see http://www.ferc.gov/industries/hydropower/indusact/hydrokinetics.asp.

[73] Remarks by Sean O'Neill, President—Ocean Renewable Energy Coalition, at the 14th Annual Congressional Renewable Energy & Energy Efficiency EXPO + Forum, June 16, 2011. DOE maintains an on-line database of ocean and hydrokinetic projects worldwide. For more information, see http://www1.eere.energy.gov/windandhydro/hydrokinetic/default.aspx.

[74] Challenges associated with calculating LCOE for ocean and hydrokinetic electricity generation technologies include (1) unknown capital costs, (2) unknown operations and maintenance costs, (3) unknown technology performance characteristics, etc.

[75] For more information on biomass feedstock, see CRS Report R41440, *Biomass Feedstocks for Biopower: Background and Selected Issues*, by Kelsi Bracmort.

[76] Biomass resource management practices may include land utilization intensity, fertilization, using more productive and/or genetically modified crops, among others. NREL's "Billion Ton" study makes some assumptions for resource management changes needed in order to achieve that resource level. For more

information, see "Biomass as Feedstock for a Bioenergy and Bioproducts Industry: The Technical Feasibility of a Billion-Ton Annual Supply," U.S. Department of Energy and U.S. Department of Agriculture, April 2005, available at http://www1.eere.energy.gov/biomass/pdfs/final_billionton_vision_report2.pdf.

[77] Energy Information Administration, see http://www.eia.gov/totalenergy/ data/monthly/pdf/ sec7_5.pdf.

[78] "Biomass as Feedstock for a Bioenergy And Bioproducts Industry: The Technical Feasibility of a Billion-Ton Annual Supply," U.S. Department of Energy and U.S. Department of Agriculture, April 2005, available at http://www1.eere.energy.gov/biomass/ pdfs/final_billionton_vision_report2.pdf.

[79] Biomass Power Association, see http://www.usabiomass.org/. Biomass material might also be co-fired with coal in conventional coal electricity generation facilities. Biomass cofiring with coal may result in improved biomass conversion efficiencies when compared to combusting only biomass. Co-firing biomass with coal may create some technical operating issues associated with tar production and fouling of electricity generating equipment. The degree to which these technical problems might be realized is dependent on the quality of the biomass material being combusted and the percentage of biomass blended and co-fired with coal. For more information see http://www.iea.org/techno/essentials3.pdf.

[80] R. Bain, W. Amos, M. Downing, and R. Perlack, "Highlights of Biopower Technical Assessment: State of the Industry and the Technology," National Renewable Energy Laboratory and Oak Ridge National Laboratory, April 2003, available at http://www.nrel.gov/docs/fy03osti/33502.pdf.

[81] Ibid.

[82] For more information about biomass feedstock characteristics see http://www1.eere.energy. gov/biomass /feedstock_databases.html. For more information about biomass feedstock logistics see http://www1.eere. energy.gov/biomass/feedstocks_logistics.html.

[83] See, for example, National Academy of Sciences, National Research Council, *Electricity from Renewable Resources: Status, Prospects, and Impediments*, National Academies Press, 2010.

[84] Terms such as LCOE, Power Purchase Agreement (PPA), contract price, and others, are sometimes used when discussing renewable electricity economics. Each of these terms has different, sometimes multiple, definitions. For example, the Federal Energy Regulatory Commission (FERC) publishes contract prices for electricity, including renewable electricity projects. Published contract prices for wind generated electricity can range between $40/MWh and $60/MWh. Comparing these contract prices with EIA LCOE estimates ($82/MWh minimum) indicates that wind electricity is being sold for less than cost. However, FERC published contract prices may not reflect any value that the wind project might receive by selling renewable energy credits (RECs). It is important to understand what is being reflected in LCOE, PPA, and contract price values.

[85] For a detailed description of this LCOE methodology, see "The Drivers of Levelized Cost of Energy for Utility-Scale Photovoltaics," SunPower Corporation, August 14, 2008, available at http://nl.sunpowercorp.be/ downloads/SunPower_levelized_cost_of_electricity.pdf.

[86] For a description of the cash flow LCOE methodology, see P. Schwabe, S. Lensink, and M. Hand, "IEA Wind Task 26: Multi-National Case Study of the Financial Cost of Wind Energy," IEA Wind, March 2011, available at http://www.ieawind.org /IndexPage POSTINGS/IEA%20WIND%20TASK%2026% 20FULL% 20REPORT%20FINAL%203%2010%2011.pdf.

[87] NREL has calculated LCOE estimates for wind and has summarized the sensitivity of LCOE values based on different assumptions. For more information, see K. Cory and P. Schwabe, "Wind Levelized Cost of Energy: A Comparison of Technical and Financing Input Variables," National Renewable Energy Laboratory, October 2009.

[88] For a description of EIA's LCOE methodology and assumptions used for the estimates, see "Levelized Cost of New Generation Resources in the Annual Energy Outlook 2011," Energy Information Administration, December 2010, available at http://www.eia.gov/ oiaf/aeo/electricity_generation.html.

[89] Union of Concerned Scientists, 2002, The Hidden Cost of Fossil Fuels, available at http://www.ucsusa.org/clean_energy/technology_and_impacts/impacts/the-hidden-cost-of-fossil.html.

[90] One study from the Brookings Institution attempts to quantify social costs, per unit of energy produced, associated with energy production. For more information, see M. Greenstone and A. Looney, "A Strategy for America's Energy Future: Illuminating Energy's Full Costs," Brookings Institution, The Hamilton Project, May 2011, available at http://www.brookings. edu/~/media/Files/rc/papers/2011/05_energy_ greenstone_ looney/05_energy_greenstone_looney.pdf. 91 For more information, see "Social Cost of Carbon for Regulatory Impact Analysis Under Executive Order 12866," DOE, Office of Energy Efficiency and Renewable Energy, available at http://www1.eere.energy.gov/ buildings/ appliance_standards/ commercial/pdfs/sem_finalrule_appendix15a.pdf.

[92] The term "power system," for the purpose of this discussion, includes electricity generators, transmission infrastructure, and electricity consumers. For more information about the North American power system, see http://www.nerc.com/page.php?cid=1|15.

[93] "Harnessing Variable Renewables: A Guide to the Balancing Challenge," International Energy Agency, 2011.

[94] Ibid.

[95] North America has three distinct interconnections: (1) Eastern Interconnect, (2) Western Interconnect, and (3) ERCOT (Electric Reliability Council of Texas) Interconnect. Each interconnect essentially operates as an independent electrical grid system.

[96] "Eastern Wind Integration and Transmission Study," National Renewable Energy Laboratory, prepared by EnerNex Corporation, February 2011, available at http://www.nrel.gov /wind/systemsintegration/pdfs/ 2010/ewits_final_report.pdf.

[97] "Western Wind and Solar Integration Study," National Renewable Energy Laboratory, prepared by GE Energy, May 2010, available at http://www.nrel.gov/wind/ systems integration/pdfs/2010/wwsis_final_report.pdf.

[98] U.S. Department of Energy, Energy Storage Program Planning Document, http://www.oe. energy.gov/ DocumentsandMedia/OE_Energy_Storage_Program_Plan_Feburary_ 2011v3.pdf.

[99] Uma Outka, *The Renewable Energy Footprint*, Stanford Environmental Law Journal, Vol. 30, p. 241, 2011.

[100] Ibid.

[101] On July 21, 2011, the Federal Energy Regulatory Commission (FERC) issued Order No. 1000—Final Rule on Transmission Planning and Cost Allocation by Transmission Owning and Operating Public Utilities. This FERC order may result in transmission capacity access for renewables, since transmission planning must take into account federal and state public policy requirements (i.e., renewable portfolio standards, etc.).

[102] "Eastern Wind Integration and Transmission Study," National Renewable Energy Laboratory, February 2011, available at http://www.nrel.gov/wind/systemsintegration/ pdfs/2010/ewits_ final_report.pdf.

103 "20% Wind Energy by 2030: Increasing Wind Energy's Contribution to U.S. Electricity Supply," U.S. Department of Energy, Energy Efficiency and Renewable Energy, July 2008, available at http://www.nrel. gov/docs/fy08osti/41869.pdf.

[104] National Academy of Sciences, National Research Council, *Electricity from Renewable Resources: Status, Prospects, and Impediments*, National Academies Press, 2010.

[105] NIMBY = Not In My Back Yard.

[106] For more information about potential water issues associated with CSP electricity generation, see CRS Report R40631, *Water Issues of Concentrating Solar Power (CSP) Electricity in the U.S. Southwest*, by Nicole T. Carter and Richard J. Campbell.

[107] "Electricity from Renewable Source: Status, Prospects, and Impediments," Chapter 5 – Environmental Impacts of Renewable Energy, National Academy of Sciences, 2010.

[108] "A National Offshore Wind Strategy: Creating an Offshore Wind Energy Industry in the United States." U.S. Department of Energy, Energy Efficiency and Renewable Energy, February 2011, available at http://www1.eere.energy.gov/windandhydro/ pdfs/national_ offshore_wind_strategy.pdf.

[109] For more information on the commercialization "valley-of-death," see "Crossing the Valley of Death: Solutions to the next generation clean energy project financing gap," *Bloomberg New Energy Finance*, June 21, 2010.

[110] For more information on energy tax policy see CRS Report R41769, *Energy Tax Policy: Issues in the 112th Congress*, by Molly F. Sherlock and Margot L. Crandall-Hollick. For more information on tax favored finance options see CRS Report R41573, *Tax-Favored Financing for Renewable Energy Resources and Energy Efficiency*, by Molly F. Sherlock and Steven Maguire.

[111] For more information see CRS Report R40913, *Renewable Energy and Energy Efficiency Incentives: A Summary of Federal Programs*, by Lynn J. Cunningham and Beth A. Roberts.

[112] For more information regarding Section 1603 of ARRA see CRS Report R41635, *ARRA Section 1603 Grants in Lieu of Tax Credits for Renewable Energy: Overview, Analysis, and Policy Options*, by Phillip Brown and Molly F. Sherlock.

[113] For more information on state incentives for renewable energy, see the Database of State Incentives for Renewables and Efficiency at http://www.dsireusa.org/.

[114] For more information see CRS Report R41720, *Clean Energy Standard: Design Elements, State Baseline Compliance and Policy Considerations*, by Phillip Brown.

[115] One example of this scenario might be the expiration of production tax credits in 2000, 2002, and 2004. For more information, see http://www.awea.org/issues/ federal_policy/ upload/PTC_April-2011.pdf.

[116] One concept worth noting here is known as the Jevons Paradox, which indicates that as efficiency increases the amount of resources demanded will also increase, not decrease as might be expected. William Jevons, in 1865, observed that as technology improved the efficiency of coal use, consumption of coal actually increased across

several industries. Whether or not the Jevons Paradox is applicable today is debatable, with experts presenting arguments that support and refute the Jevons Paradox. A high level overview of the Jevons Paradox is available at http://en.wikipedia.org/wiki/Jevons_paradox. For more information, see CRS Report RL31188, *Energy Efficiency and the Rebound Effect*, by Frank Gottron.

[117] Energy Information Administration, http://www.eia.gov/cfapps/ipdbproject/ IEDIndex3.cfm? tid=92&pid=46&aid=2.

[118] Rich Brown, Sam Borgeson, Jon Koomey, Peter Biermayer, "U.S. Building-Sector Energy Efficiency Potential", Environmental Energy Technologies Division, Ernest Orlando Lawrence Berkeley National Laboratory, Report LBNL-1096E, September 2008.

[119] U.S. Department of Energy, Office of Energy Efficiency and Renewable Energy, http://www.energy.gov/energyefficiency/index.htm, and the U.S. Environmental Protection Agency, http://www.energystar.gov/.

[120] National Renewable Energy Laboratory, http://www.nrel.gov/learning/re_biofuels.html.

[121] Energy Information Administration, http://www.eia.gov/renewable/alternative_transport_ vehicles/pdf/ afvatf 2009. Pdf

[122] U.S. energy use at the national scale is measured in quadrillion British thermal units (Btu), or "quads."

[123] REN21. 2011. Renewables 2011 Global Status Report (Paris: REN21 Secretariat), http://www.ren21.net/REN21 Activities/Publications/GlobalStatusReport/GSR2011/tabid/56142/Default.aspx.

[124] Energy Information Administration, Annual Energy Outlook 2011, http://www.eia.gov/ forecasts/aeo/pdf/0383(2011).pdf.

[125] Eckhart, M. "Renewable Energy Exceeds 50 GW and Enters Decade of Scale-Up," Infrastructure Solutions Magazine, April 2011, available at http://www.acore.org/wp-content/uploads/2011/04/Infrastructure-Magazine-Article-V4.pdf.

[126] Energy Information Administration, "Annual Energy Outlook 2011: with projections to 2035," DOE/EIA-0383(2011), April 2011, available at http://www.eia.gov/forecasts/aeo/ pdf/0383(2011).pdf.

In: The Transformation of Electrical Power: Key Issues
Editor: Dawson B. Munoz
ISBN: 978-1-62618-619-4

Chapter 4

THE SMART GRID AND CYBERSECURITY—REGULATORY POLICY AND ISSUES*

Richard J. Campbell

SUMMARY

Electricity is vital to the commerce and daily functioning of United States. The modernization of the grid to accommodate today's uses is leading to the incorporation of information processing capabilities for power system controls and operations monitoring. The "Smart Grid" is the name given to the evolving electric power network as new information technology systems and capabilities are incorporated. While these new components may add to the ability to control power flows and enhance the efficiency of grid operations, they also potentially increase the susceptibility of the grid to cyber (i.e., computer-related) attack since they are built around microprocessor devices whose basic functions are controlled by software programming. The potential for a major disruption or widespread damage to the nation's power system from a large scale cyberattack has increased focus on the cybersecurity of the Smart Grid.

Federal efforts to enhance the cybersecurity of the electrical grid were emphasized with the recognition of cybersecurity as a critical issue for electric utilities in developing the Smart Grid. The Federal Energy Regulatory Commission (FERC) received primary responsibility for the reliability of the bulk power system from the Energy Policy Act of 2005. FERC subsequently designated the North American Electric Reliability Corporation (NERC) as the "Electric Reliability Organization" (ERO) with the responsibility of establishing and enforcing reliability standards. Compliance with reliability standards for electric utilities thus changed from a voluntary, peer-driven undertaking to a mandatory function. The Energy Independence and Security Act of 2007 (EISA) later added requirements for "a reliable and secure electricity infrastructure" with regard to Smart Grid development. NERC is also responsible for standards for critical infrastructure protection (CIP) which focus on planning

* This is an edited, reformatted and augmented version of a Congressional Research Service publication, CRS Report for Congress R41886, from www.crs.gov, dated June 15, 2011.

and procedures for the physical security of the grid. Self-determination is a key part of the CIP reliability process. Utilities are allowed to self-identify what they see as "critical assets" under NERC regulations. Only "critical cyber assets" (i.e., as essential to the reliable operation of critical assets) are subject to CIP standards. FERC has directed NERC to revise the standards so that some oversight of the identification process for critical cyber assets was provided, but any revision is again subject to stakeholder approval. While reliability standards are mandatory, the ERO process for developing regulations is somewhat unusual in that the regulations are essentially being established by the entities who are being regulated. This may potentially be a conflict of interest, especially when cost of compliance is a concern, and acceptable standards may conceivably result from the option with the lowest costs. Since utility systems are interconnected in many ways, the system with the least protected network potentially provides the weakest point of access.

Cybersecurity threats represent a constantly moving and increasing target for mitigation activities and mitigation efforts could likewise spiral upward in costs. Recovery of costs may present a major challenge especially to distribution utilities and state commissions charged with overseeing utility costs. EISA only requires states to consider recovery of costs related to Smart Grid systems. FERC has jurisdiction over the bulk power grid, and cannot compel entities involved in distribution to comply with its regulations. Recoverability from a cyber attack on the scale of something which could take down a significant portion of the grid will likely be very difficult, but maintaining a ready inventory of critical spare parts in close proximity to key installations could quicken recovery efforts from some types of attack. The electricity grid is connected to (and largely dependent on) the natural gas pipeline, water supply, and telecommunications systems. Technologies being developed for use by the Smart Grid could also be used by these industries. Consideration could be given to applying similar control system device and system safeguards to these other critical utility systems.

INTRODUCTION

Electricity is vital to the commerce and daily functioning of United States. The electrical grid of the United States consists of all the power plants generating electricity, together with the transmission and distribution lines and their associated transformers and substations which bring power to end-use customers. Electric power generation in the United States is currently dominated by the use of combustible fuels, such as coal and natural gas, or from biomass. These fuels are burned to produce steam in boilers which is used to turn turbine-generators which produce electricity. Nuclear power uses heat from the decay of radioactive elements to produce steam. However, electricity can also be generated directly by wind turbines, solar power, geothermal energy, and hydropower. Figure 1 illustrates the phases of the electric power production process. Electricity from large central station power plants is routed to a step-up transformer which raises the voltage so the power can be sent over high-voltage transmission lines to step-down transformers in substations which then transfer this energy at lower voltages via distribution lines to industrial, commercial, and residential users.

Generally, electricity must be used as soon as it is produced because large amounts of electricity cannot be easily stored.

Originally, the individual company systems were not linked, but with greater electricity demand came the necessity of sharing generation resources. This sharing of generation resources required an interconnection of separate company systems to enable power sales and transfers. These aggregated power systems form three major "interconnections"—the Eastern and Western interconnections, and the Electric Reliability Council of Texas which includes most of the state of Texas.

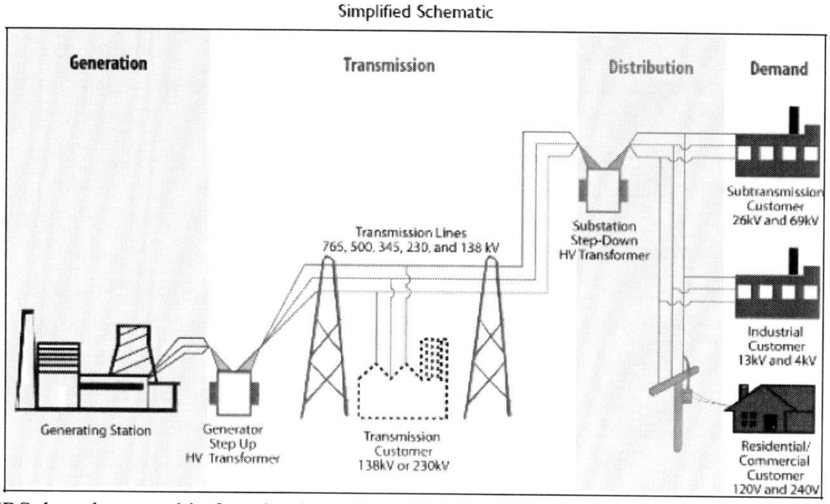

Source: CRS, based on graphic found at https://reports.energy (p. 5).

Figure 1. Elements of the Electric Power System.

Within these interconnections are reliability regions, and a number of balancing authorities[1] which "integrate resource plans ahead of time, maintain load-interchange generation balance" within a balancing authority area, and "support interconnection frequency in real-time."[2] The grid also connects the many publicly and privately owned electric utility and power companies in different states and regions of the United States[3] (and in Canada and Mexico).

Congress passed the Public Utility Regulatory Policies Act of 1978 (P.L. 95-617) which allowed non-utility power producers into wholesale power markets (e.g., for the sale of electricity to entities other than the end-user of power). The passage of the Energy Policy Act of 1992 (P.L. 102-486) served to further promote greater competition in the bulk power markets.[4] As a result, in many parts of the United States, the electric power industry began to transition from highly regulated, local monopoly companies which generated, transmitted, and distributed electricity to end-use customers, to a business in which power generation is competitive while the industry's transmission and distribution functions are still highly regulated. While the federal government regulates the electric power transmission function and wholesale power markets, regulation of the distribution function of the electric power business is still largely carried out by state government agencies.

Much of the infrastructure which serves the U.S. power grid is aging. The average age of power plants is now over 30 years; most of these facilities were originally designed to last 40 years.[5] Electric transmission and distribution system components are similarly aging, with power transformers averaging over 40 years of age,[6] and 70% of transmission lines being 25 years or older.[7] As components of the system are retired, they are replaced with newer components often linked to communications or automated systems.

With changes in federal law,[8] regulatory changes, and the aging of the electric power infrastructure as drivers, the grid is changing from a largely patchwork system built to serve the needs of individual electric utility companies to essentially a national interconnected system capable of accommodating massive transfers of electrical energy between regions of the United States. The modernization of the grid to accommodate today's power flows, serve reliability needs, and meet future projected uses is leading to the incorporation of information processing capabilities for power system controls and operations monitoring. The "Smart Grid" is the name given to the evolving electric power network as new information technology systems and capabilities are incorporated. While these new components may add to the ability to control power flows and enhance the efficiency of grid operations, they also potentially increase the susceptibility of the grid to cyber (i.e., computer-related) attack since they are built around microprocessor devices whose basic functions are controlled by software programming and subject to manipulation over a network.

The information processing attributes which make the smarter grid attractive are the very same attributes which can increase the vulnerability of the electric power system and its critical infrastructure to a potentially devastating cyber attack. The potential of the Smart Grid to revolutionize the ways power is generated and used is great, but so too are the potential risks if not sufficiently addressed.

The Energy Independence and Security Act of 2007[9] (EISA) (P.L. 110-140) defined the attributes of a Smart Grid and plans for its development. The potential for a major disruption or widespread damage to the nation's power system from a large scale cyberattack has increased focus on the cybersecurity of the Smart Grid. While a nation-state might have the resources and capability to create widescale disruption of the U.S. grid (to create economic disorder or as part of a physical attack), terrorists or hackers are less likely to have the sophistication or resources necessary to inflict damage on a national scale. But the possibility of collaboration between terrorists and hackers could increase the capability of these groups to launch network-based cyberattacks against critical U.S. infrastructure.[10]

Among the issues for Congress is whether the existing requirements of current law adequately assure the cybersecurity and reliable operation of the U.S. electricity grid today (and the Smart Grid of tomorrow). This report will look at issues key to these determinations, discussing the scope and point of application of regulations, and where the overall responsibility rests for the cybersecurity of the grid.

ATTRIBUTES OF A SMART GRID

The U.S. Department of Energy (DOE) summarized its view of the potential of the Smart Grid by the year 2030 as:

> ... a fully automated power delivery network that monitors and controls every customer and node, ensuring a two-way flow of electricity and information between the power plant and the appliance, and all points in between. Its distributed intelligence, coupled with broadband communications and automated control systems, enables real-

time market transactions and seamless interfaces among people, buildings, industrial plants, generation facilities, and the electric network.[11]

A "smarter" grid is gradually being introduced as upgrades to the systems and components of the power generation, transmission, and distribution infrastructure. Substation and distribution substations are being automated with superior switching capabilities to enhance current flows and control of the grid. Devices called "phasor measurement units" are also being added to substations to make time and location-specific measurements of transmission line voltage, current, and frequency (i.e., synchrophasor measurements made 30 times per second instead of data measured once every two to four seconds by current industrial control systems) providing better tools to improve power system reliability.[12] DOE's strategic plan targets the deployment of 1,000 phasor measurement units by 2013.[13]

Advanced meters are being introduced in some utility markets which can offer customers the ability to leverage demand response[14] regimes and technologies, and potentially benefit from the interactive features of the Smart Grid by shifting their energy use to off-peak hours. Expectations vary of what a Smart Grid could accomplish, and the estimated costs of a system rise with the increased scope and attributes of a system. Some see the Smart Grid as only an enhancement of today's existing transmission and distribution systems focused on improving the reliability and efficiency of the grid. Others see the Smart Grid of the future as an integrated system spanning the nation from coast to coast, able to seamlessly combine distributed resources and central power stations across the three major interconnections[15] of the U.S. grid. Under such a vision, distributed and renewable energy resources could be efficiently integrated into the grid, with power from (for example) intermittent wind generation channeled by sensors and intelligent electronics from multiple widely dispersed sites to where power is needed anywhere on the grid.

Figure 2 shows a possible configuration of a Smart Grid system illustrating components of an interconnected, two-way intelligent network, and distributed and central station sources of power.

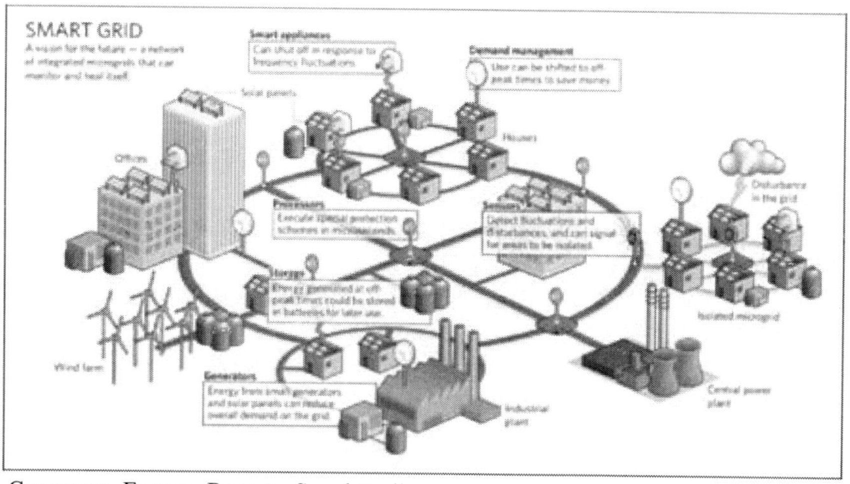

Source: Consumer Energy Report. See http://www.consumerenergyreport.com/wp-content/uploads /2010/04/ smartgrid.jpg.

Figure 2. One Conceptualization of the Smart Grid.

The efficiency and economy of all grid operations could conceivably be optimized by similarly harnessing all power generation to take advantage of a wide range of generation and storage resources across the United States.

Defining Cybersecurity for the Smart Grid

A prolonged disruption of energy from the grid was described in a report for the Presidential Commission on Critical Infrastructure[16] as an event which would seriously affect every infrastructure, with the possibility of a cyberattack causing such damage considered as a possibility.

> The widespread and increasing use of Supervisory Control and Data Acquisition17 (SCADA) systems for control of energy systems provides increasing ability to cause serious damage and disruption by cyber means. The exponential growth of information system networks that interconnect the business, administrative, and operational systems contributes to system vulnerability.[18]

EISA gave the National Institute of Standards and Technology (NIST) the role of coordinating the development of a framework to enable the development of the Smart Grid in a safe and secure manner, and NIST issued its first guidelines for cybersecurity of the Smart Grid in 2010. Because cybersecurity threats were perceived as "diverse and evolving," NIST advocated a defense-indepth strategy with multiple levels of security because no single security measure could counter all types of threats.[19]

The key to NIST's suggested approach is the determination of risk (i.e., the potential for an unwanted outcome resulting from internal or external factors, as determined from the likelihood of occurrence and the associated consequences) as quantified by the threat (e.g., event, actor or action with potential to do harm), the vulnerability (e.g., weakness in the system), and the consequences (e.g., physical impacts) to the system.[20]

Cybersecurity Vulnerabilities of the Smart Grid

A smarter grid is largely the result of the modernization of the grid. Most sectors of business and industry have gone from the analog age of knobs and dials to the digital age of sensors and electronic displays in the recent past, and that process is now reaching into the nation's electric power infrastructure.

Part of the delay has been due to the long-lived nature of the capital assets which make up the industry. The power plants and other expensive components of the grid can function for many productive years if maintenance of the systems is kept up. However, much of the delay in modernization has been due to cost concerns, as many electric utilities seek to manage expenses by delaying replacement of aging systems as long as possible. Several of the better-known vulnerabilities of a smarter grid are discussed in the following paragraphs.

Control Systems

Industrial control (IC) systems are particularly vulnerable to cyberattack because of their intelligence and communications capabilities. IC systems perform a number of functions in the electrical grid, ranging from microprocessor-based control systems which control the actuation and operation of one or more devices, to more sophisticated industrial IC systems which can manage entire industrial processes or automated systems. SCADA systems are a well-known application of remote IC used to monitor and control electric transmission system components.

While cyber-intrusions into the U.S. grid have been reported in recent years,[21] no impairment or other damage has been publicly reported as a result of the attacks. By exploiting loopholes in cybersecurity, cyberattackers could breach the privacy of customers power usage data and access large numbers of meters, perhaps sending deliberately misleading information to the grid. This could potentially overload systems or cause grid operators to respond to false readings. However, concerns exist as to the potential damage that could result in the future from malware left behind by such intrusions or doorways created in systems which could be exploited.[22] The revelation of the complexities of the Stuxnet worm[23] and the alleged targeting of the control systems of a nuclear power plant in Iran[24] have raised additional concerns about the vulnerability of electric power systems worldwide.

The Stuxnet worm reportedly affected the logical decision making functions of the control systems of one particular type of manufacturing facility. Theoretically, Stuxnet-like malware could be adapted to impair other types of process or control systems in malicious or unpredictable ways.

Malware modified in such ways could overload and damage targeted equipment and systems in critical infrastructure.[25] The Department of Homeland Security (DHS) conducted a simulated cyberattack on an electric generator control room and partially destroyed a large diesel-electric generator secured for the test as a demonstration of existing power system vulnerabilities.[26]

Connections through Other Systems

Utilities have routinely accessed control systems either wirelessly or by telephone lines for many years. In fact, telephone lines often used to access corporate systems may even provide access to utility control systems. A recent GAO report highlighted vulnerabilities at the Tennessee Valley Authority, a large federal utility in the United States.

> ... the interconnections between TVA's control system networks and its corporate network increased the risk that security weaknesses, on the corporate network could affect control systems networks and we determined that the control systems were at increased risk of unauthorized modification or disruption by both internal and external threats.[27]

Communications and Internet Access

Grid devices capable of two-way communications are considered to be potential points of unauthorized system access, and can represent a potential cybersecurity vulnerability. While security protocols may exist to prevent unauthorized entry, wireless networks can be monitored and potentially hacked by cybercriminals. Smart meters are another example of new applications in which the security of data has been mentioned as a concern. At the heart of smart meters are semiconductor chips which allow data about energy use to be sent wirelessly to the electric distribution company, and which potentially allows the meter to control the flow of power to customers. The security of the encrypted information in the communications protocol used by many of these devices has been questioned in the past, and questions have been raised about the effectiveness of efforts to patch identified flaws. The primary consideration in the choice of wireless protocol by semiconductor providers seems to be cost of the system, a criteria which is not always compatible with cybersecurity goals for secure communications.[28] While the developing Smart Grid is likely to use today's internet and wireless communications systems for many years to come, security concerns and data usage requirements may move the system to dedicated communications and information channels serving uniquely Smart Grid uses.

Potential access to SCADA systems via the Internet has often been cited as a key vulnerability for electric power systems. Hackers have used the Internet to launch denial of service,[29] phishing,[30] and various other forms of cyberattack. The effective recording, processing, and exchanging of data is becoming increasingly critical to the reliability of the power system,[31] and deliberately introduced misinformation can be damaging. Internet-linked communications systems may be important to today's interconnected grid, but the Internet also provides a ready path to cyberattack from any corner of the world wide web.

Combined Cyber- and Physical Attacks

The vulnerability of key components of the grid to a physical attack has been long known, but a coordinated and sustained physical and/or cyberattack[32] could be problematic. After a cyberattack, restored systems may be repaired and protected against known vulnerabilities, but a coordinated cyberattack over a short time period targeting different Smart Grid systems could slow recovery. The same would hold true for repeated physical attacks, or combined cyber- and physical attacks on key grid components. Under such circumstances, the resources of emergency programs in place to replace important grid systems and components (such as the Edison Electric Institute's "Spare Transformer Equipment Program"[33]) may not be adequate to provide enough replacement parts to fully restore reliable grid operations in a timely manner.

Supply Chain

The components of smarter grid devices present another potential vulnerability concern. Most of the smart meter, sensor, and other equipment makers are international companies who obtain their components from international sources. Taiwan, Singapore, China, and

South Korea are among the largest overseas manufacturers of semiconductors and microprocessors for smart devices. The reliable operation of semiconductor and microprocessor-based devices is based on the low-level firmware[34] controlling the device's basic functions. Firmware in the form of fixed machine-language binary code is found in almost all the electronic devices making up smarter grid products such as programmable controllers and programmable logic arrays. If a hacker or cybercriminal gained access to such devices (especially during the manufacturing process), a section of code could be covertly inserted in the device and activated in such a way as to impair its functioning in a reliable manner. Some might suggest random or statistically based testing of the firmware in smarter grid devices. But the impairment would not need to be placed in all such devices coming off an assembly line. If a large enough sample was impaired, the effect might be enough to cast doubt on the reliability of a whole class of such devices.

Geomagnetic Impairment

An electromagnetic pulse (EMP) is a large burst of electromagnetic radiation resulting from the detonation of a nuclear weapon or from other means. EMPs can cause voltage and current surges which can severely damage electrical equipment and systems. A high-altitude nuclear explosion could cause an EMP which could severely impair an electric power system not shielded from the effects of an EMP.

However, deliberate, large-scale, man-made attacks are not the only potential threat which could take down a substantial portion of the electric grid. Radiation or charged particles ejected into space weather from solar flare eruptions could cause geomagnetic storms which could interfere with telecommunications networks or damage orbiting satellites. Periodically, these eruptions from the sun are powerful enough to disrupt the operations of electric power systems. Solar flares occur in cycles of almost 11 years. [35] Occasionally, severe space weather results in a huge eruption of particles which can be particularly disruptive to electric power systems, and induce ground currents powerful enough to actually melt the copper windings of electric power transformers. The last major solar flare eruption in 1989 caused blackouts across the Canadian province of Quebec for nine hours. But even greater storms can occur perhaps every 100 or more years. The first major eruption of this size was observed in 1859. It destroyed much of the world's telegraph infrastructure of the time. Another event observed in 1921 demonstrates that although these events are rare, they are likely to occur again and could have tremendous societal and economic impacts for the United States:

> ... the occurrence today of an event like the 1921 storm would result in large-scale blackouts affecting more than 130 million people and would expose more than 350 transformers to the risk of permanent damage.[36]

It could take weeks or months to replace most of the damaged transformers. Some of the larger units could take years to replace due to the fact there are no U.S. power transformer manufacturers,[37] and a multi-year backlog exists for the larger units.

CURRENT CYBERSECURITY POLICY FOR THE SMART GRID

A smarter grid is increasing reliance on automated systems to manage energy sources, power control devices, system information, and power flows. Automation will allow information from smarter processors and sensors to be processed at much faster rates, giving power system operators more accurate and timely information, and giving customers greater input into their own power use. But, the possibility that such systems could be highjacked in such a way as to bring down the grid regionally or even nationally, and cause long-term damage to critical systems and components, is giving Congress reason to consider the cybersecurity of the electric power infrastructure.

The passage of the Homeland Security Act of 2002 (P.L. 107-296) (HSA) established the Department of Homeland Security. HSA increased penalties for certain computer-based crimes, and in addition to improving on the earlier Computer Fraud and Abuse Act of 1986,[38] made critical infrastructure protection a key mission of DHS.

Federal efforts to enhance the cybersecurity of the electrical grid were reemphasized with the recognition of cybersecurity as a critical issue for electric utilities in developing the Smart Grid. The Energy Independence and Security Act (EISA) of 2007 (P.L. 110-140) thus added requirements for "a reliable and secure electricity infrastructure" with regard to Smart Grid development. EISA directed NIST to develop a framework for protocols and standards for the Smart Grid to achieve "interoperability"[39] of devices and systems. The American Recovery and Reinvestment Act of 2009 (P.L. 111-5) provided NIST with funds to proceed with this mandate.

Status of Regulatory Implementation

NIST issued a report, "NIST Framework and Roadmap for Smart Grid Interoperability Standards" (NIST SP 1108, January 2010), listing 75 existing standards which are likely to be applicable for an interoperable Smart Grid, and identified 15 high-priority gaps and harmonization issues. NIST also issued a "Smart Grid Cybersecurity Strategy and Requirements" report (NISTIR 7628, September 2010) assessing requirements to ensure the security and reliability of a modernized grid. This report also classified Smart Grid interfaces according to potential impacts that could result from compromise of these systems.

The Federal Energy Regulatory Commission (FERC or the Commission) received primary responsibility for the reliability of the bulk power[40] system from the Energy Policy Act of 2005 (EPACT05) (P.L. 109-58). EPACT05 also gave FERC the authority to commission an "Electric Reliability Organization" (ERO) to establish and enforce reliability standards. FERC subsequently designated the North American Electric Reliability Corporation (NERC) as the ERO. Compliance with reliability standards for electric utilities thus changed from a voluntary, peer-driven undertaking to a mandatory function. While NERC is responsible for standards for the bulk power system in the United States, it is also seeking recognition by applicable governmental authorities in Canada and Mexico to establish and coordinate reliability standards for the bulk power systems of these respective countries.[41] NERC is also responsible for standards for critical infrastructure protection (CIP) which focus on planning and procedures for the physical security of the grid. NERC has delegated much of the daily process of maintaining reliability of the grid to regional entities via delegation agreements which are subject to FERC approval. Compliance and enforcement with NERC standards is therefore largely a responsibility of these regional entities, which are mostly responsible for imposing penalties for violation of reliability standards.[42] NERC's CIP standards essentially approach cybersecurity as an extension of reliability, and address nine areas including *Sabotage Reporting*, *Critical Cyber Asset Identification*, and *Electronic Security Perimeters*, and *Recovery Plans for Critical Cyber Assets*.[43] These standards identify what utilities must do to protect "cyber assets"[44] seen as critical to the reliability of the bulk power[45] system.

EISA directed FERC to work with NIST in developing Smart Grid interoperability standards, and required FERC to institute a formal rulemaking process to adopt interoperability (including cybersecurity) standards when FERC was satisfied that a consensus on defining the standards had been reached. NIST advised FERC in October 2010 that it had identified five "foundational" sets of standards[46] which may allow FERC to go forward with instituting rulemaking proceedings to adopt standards to ensure Smart Grid functionality, interoperability, and cybersecurity. The standards would be updated as Smart Grid requirements and technologies evolve.

FERC held a technical conference in January 2011 to determine if there was "sufficient consensus" that the five sets of standards proposed by NIST were adequate for standards and protocols to be adopted. The comments of conference participants showed that a consensus on the standards had not in fact been reached with several participants stating that they were unsure of what would constitute a consensus, and questioning whether such standards would then automatically become "mandatory and enforceable." FERC issued a supplemental notice in February 2011 (requesting all comments by April 2011) to aid the Commission's determination as to whether a consensus exists on the proposed NIST interoperability standards. Additional questions were asked in the notice to solicit opinions on related topics including the NIST process for determining the readiness of a standard with regards to interoperability and cybersecurity, whether making standards enforceable would achieve interoperability goals, how testing and certification of cybersecurity requirements could be incorporated into the standards process, and whether the Commission should consider different criteria for evaluating interoperability and functionality (and the extent to which cybersecurity is an element of each).[47]

Assessment of Cybersecurity Progress by Government Accountability Office

Congress asked the U.S. Government Accountability Office (GAO) to assess progress by NIST and FERC on Smart Grid cybersecurity guidelines and standards. GAO issued a report with its findings in January 2011.[48] GAO found that NIST has largely addressed "key cybersecurity elements" such as the cybersecurity risks of Smart Grid systems and had identified security controls essential to such systems. GAO also found that NIST had not addressed the risks of attacks on Smart Grid systems using both cyber- and physical means. GAO recommended that NIST finalize its plan and schedule for updating its cybersecurity guidelines to include these elements.

GAO also pointed out that while EISA had given FERC authority to adopt Smart Grid standards, it did not give FERC specific enforcement authority over implementation of standards. GAO recognized that a regulatory divide exists between federal, state, and local entities on various aspects of Smart Grid interoperability and cybersecurity. As such, GAO states that such standards will remain voluntary unless regulators use other authorities to enforce standards compliance. GAO recommended that FERC develop a coordinated approach (with other regulatory jurisdictions) to monitor voluntary standards and address any gaps in compliance.

FERC and NIST agreed with GAO's recommendations.[49] However, FERC Chairman Wellinghoff stated that while a coordinated approach would be pursued, FERC may lack the authority to address any gaps in compliance, and that this being the case, Congress may want to consider the matter further. Furthermore, in addressing GAO's comments, the Chairman broached the question of whether other-than-voluntary compliance with Smart Grid standards by "relevant manufacturers and utilities" is what Congress intended in EISA.[50]

Meanwhile, NIST has communicated to FERC its own concerns that under the existing process, it could take years for the Commission to adopt rules on the many individual standards needed to achieve interoperability of the Smart Grid while recognizing that many "customized" upgrades to the grid continue to be installed. NIST suggested that in the absence of formal standards, FERC could request "interoperability roadmaps" from utilities to determine if they addressed interoperability based on the NIST framework in the utilities' grid modernization efforts.[51]

Assessment by DOE's Inspector General of FERC's Monitoring of Grid Cybersecurity

The DOE Inspector General released an audit report[52] in January 2011 of FERC's monitoring of grid cybersecurity, with regard to the Commission's responsibilities under EPACT05. The report found that CIP cybersecurity standards developed "did not include a number of security controls commonly recommended for government and industry systems," and criticized FERC's oversight of the process for developing these standards, citing a need for FERC to use its existing authority to ensure timely standards development to address emerging security threats. The report also said that FERC's implementation approach and schedule for CIP standards did not adequately consider risks to information systems, since FERC was focusing on documentation of controls rather than implementation of controls. FERC was advised to revise its focus to ensure that controls to address higher threat risks are given priority. The report noted that these problems existed, in part, because the Commission had "only limited authority" to ensure cybersecurity over the bulk electric system, and could not implement its own reliability standards or issue alerts. However, the report went on to conclude that even "when such authority did exist," FERC did not always act "to ensure that cybersecurity standards were adequate."

FERC Chairman Wellinghoff mostly concurred with the report's recommendations[53] while noting that "the current statutory framework is adequate for addressing emerging cybersecurity threats through mandatory standards." The Chairman stated that FERC will make every effort "to approve CIP standards as soon as possible after they are developed by NERC."[54]

Policy Concerns

Self-determination is a key part of the CIP reliability process. Utilities are allowed to self-identify what they see as "critical assets"[55] under NERC regulations. While compliance with CIP standards became mandatory in 2010, FERC mandated modifications to the criteria defining critical assets for reliability purposes. Only "critical cyber-assets" (i.e., as essential to the reliable operation of critical assets) are subject to CIP standards. FERC directed NERC to revise the standards so that some oversight of the identification process for critical cyber-assets was provided, but any revision is again subject to stakeholder approval. Under current law, FERC cannot issue its own reliability standards.

While reliability standards are mandatory, the ERO process for developing regulations is somewhat unusual in that the regulations are essentially being established by the entities who are being regulated.[56] This can potentially be an issue when cost of compliance is a concern, and acceptable standards may conceivably result from the option with the lowest costs. While FERC ultimately has approval authority over the regulations NERC submits and can remand such regulations it judges as not satisfying requirements, any such revisions are ultimately subject to NERC stakeholder approval. NERC standards could therefore be seen as a minimum threshold for compliance, as some utilities may choose to go beyond what is minimally required. But since utility systems are interconnected in many ways, the system with the least protected network potentially provides the weakest point of access.

Other Issues in Federal Smart Grid Cybersecurity Policy

Legacy Systems

Legacy devices and systems (such as many SCADA and substation automation systems) may represent as much of a vulnerability to cybersecurity as new Smart Grid components. They were not designed with cybersecurity in mind, and are often interconnected either via the Internet or by other, sometimes "unsecured" avenues. DOE describes the legacy system communications issue as follows:

> The legacy communication methods that now support grid operations also provide potential cyber attack paths. Many cyber vulnerabilities have been identified by cyber security assessments of SCADA systems. Power grid substation automation and security have also recently been evaluated. The level of automation in substations is increasing, which can lead to more cyber security issues. The cyber security issues identified during the assessments and evaluations need to be resolved, but the known issues should not be construed as a complete assessment of the current power grid security posture.[57]

Some legacy grid devices are being retrofitted with communications capabilities to allow them functionality in the smarter grid, or permit easier maintenance, potentially introducing grid vulnerabilities which may not have existed before. NIST has suggested the establishment of "Priority Actions Plans" to resolve interoperability issues between legacy equipment and Smart Grid systems as these are installed.[58]

FERC specifically rejected proposals by commenters for additional interoperability standards (at this time) addressing existing resources or equipment and cost-effective integration of legacy equipment, and interfaces between utilities.[59] Replacement of these systems is recognized by FERC as a possible option for utilities to apply for recovery of the "stranded costs" of legacy systems replaced by Smart Grid investments.[60]

Leadership of Federal Authority for Smart Grid Cybersecurity

The issue of which, or whether, a single federal agency should have ultimate authority over Smart Grid cybersecurity remains largely unresolved. DHS, DOE, and FERC have all either stated claims or been mentioned as candidates for overall leadership of the issue based upon implications to national security, energy policy and technology leadership, or industry understanding and organizational mission, but the current missions of these agencies is only the starting point for a discussion. Since Smart Grid devices are being installed beyond the federally regulated power grid, the system could potentially face intrusions from cyberattacks initiated on less-protected distribution systems. Leadership of Smart Grid and cybersecurity issues may require the ability to communicate issues and push solutions in forums beyond federal jurisdictions if the goal is to protect against cyber-threats to the grid for the entire United States.

DHS's mission to protect national security against cyber-threats currently centers on "civilian government systems" and DHS works with industry to "secure critical infrastructure and information systems." DHS is focused on analyzing cyber-threats and reducing vulnerabilities, and distributing information on potential cyberattacks to ensure the safety of cybersystems.[61] DHS will distribute warnings of cyber-threats to other federal agencies, state and local governments and to industry.

DOE has the technical and research capabilities of the national laboratories to draw upon with regard to cybersecurity and the Smart Grid, as well as its own internal policy offices which deal with an array of electricity system issues. DOE has worked with electric utilities to understand the cyber-vulnerabilities of the grid, and has tasked the national laboratories with helping to develop technologies and software patches to identify problems and harden legacy control systems against potential cyber-intrusions.

FERC is an independent, self-funded regulatory agency within DOE, with statutory authority over major aspects of the wholesale electric, natural gas, hydroelectric, and oil pipeline industries.[62] As such, FERC's responsibility for the reliability of the grid and related ratemaking authority allow the Commission to discover issues and authorize utilities to recover the jurisdictional portion of costs related to grid critical infrastructure protection upgrades. FERC recognizes that EISA does not make Smart Grid standards mandatory, nor does EISA give it authority to make or enforce such standards.[63] FERC has jurisdiction over the bulk power system, but this currently excludes authority over the nation's electric distribution systems where much of the smarter grid will be deployed. The Commission appears to view its reliability mandate under EISA as giving it the authority to adopt Smart Grid standards which will be applicable to all electric power facilities and devices with Smart Grid features, including those at the distribution level. FERC further ties cybersecurity to reliability by proposing a "harmonization" of each Smart Grid protocol and reliability standard developed. [64]

Cost Challenges of Maintaining a Cyber-Secure Smart Grid

Cybersecurity threats present a constantly moving and growing need for mitigation activities. As new devices and systems are being developed for the Smart Grid, so too is the array of infrastructure elements which must potentially be protected. This is a key concern for utilities, as the resources which may have to be applied to cybersecurity concerns could likewise spiral upward in growth and costs. Utilities may understand how a smarter grid can enhance their efficiency and reliability, and the need for cybersecurity safeguards. Commercial and industrial customers may also see potential benefits in Smart Grid-enabled programs (such as demand response), and it is likely they will fund a significant portion of smarter grid improvements through rates. But at present very few consumers at the residential level seem to understand the potential of the Smart Grid, or have reason to believe it can do anything for them except increase rates.[65] One recent study expects cybersecurity spending to represent approximately 15% of total Smart Grid capital investment between 2010 and 2015 (with total investment estimated to be $1.5 billion in North America).[66]

Smart Grid enhancements across the United States may vary by utility, state, region, and regulatory jurisdiction. Recovery of costs may present a major challenge especially to distribution utilities and state commissions charged with overseeing utility costs. When faced with such decisions, the perception of the potential risks to the utility's local system will likely bear on what costs the utility's ratepayers should pay. EISA requires states only to consider recovery of costs related to Smart Grid systems. FERC has jurisdiction over the bulk power grid, i.e., transmission and wholesale rates in interstate commerce. FERC cannot compel entities involved in distribution to comply with its regulations. Since distribution utilities are regulated by state or local public utility commissions, Smart Grid investments will likely require approval by these jurisdictions. The issue of whether Smart Grid standards for interoperability should be voluntary or mandatory for these jurisdictions remains unresolved.

Privacy and Data Security

The sharing of information in applications used by the smarter grid has raised questions on the safety of that information. Security of customer information in wireless applications, and how personal data characteristics (such as customer usage information) can be protected are issues often mentioned in discussions of the Smart Grid and cybersecurity. Encryption of data (with limited decryption for data checking), and aggregation of data at high levels to mask individual usership have been mentioned as ways to protect the identity of individual customers. Limiting the amount of data to just information needed for billing purposes has been suggested as a remedy to the potential for misuse of data.[67] Real-time monitoring of these networks may alert system operators to suspicious activities. But even these methods might not be enough to guard against a sophisticated intruder if data security is taken for granted. The development of Smart Grid standards for electric utilities and their partners governing the collection and use of customer data may be a possible next step.

If customers participate in demand-side management programs, then customer usage data can provide a wealth of information for a variety of programs for interruptible loads or time-of-use rates. But customer-specific data stored in home area networks (HANs), or customer-specific data communicated between the HAN and distribution utility (or load aggregator) must be secure to protect the privacy of information. The advent of electric vehicles may offer another potential payload of data on customer movement and habits, if data collected or stored is not restricted to electricity consumption for billing purposes.

The National Regulatory Research Institute recommended in a recent report[68] that public utility commissions should define the information which utilities will collect, determining with whom and for what purpose it should be shared, and assess the need for protecting the data. While allowing states to develop their own policies may be accomplished more quickly, the report also advocates for a national approach to Smart Grid privacy issues.

DOE has worked with electric utilities and the Electric Power Research Institute (EPRI) to develop profiles for Smart Grid products to address concerns about data privacy. These profiles are meant to provide guidance on how to build cybersecurity into Smart Grid applications under development, so that they can be more securely integrated into the Smart Grid.[69]

Government-Industry Coordination on Threats

In 2010, the Edison Electric Institute (EEI) issued its own "Principles for Cyber Security and Critical Infrastructure Protection"[70] to discuss development of a framework for addressing cybersecurity threats. EEI, the trade association of U.S. shareholder-owned electric utilities, suggested a coordinated effort of government, industry, and suppliers of critical grid systems and components was necessary to protect the grid from cyberattacks as a "shared responsibility." EEI believes that the sharing of intelligence on evolving threats and a clear plan of action on imminent threats is key to this arrangement. EEI also expects that such a policy would help protect national security and the public welfare, and aid in the prioritization of assets which need enhanced security. EEI expects that such a partnership could "utilize all stakeholder's expertise," allowing owners and operators of critical infrastructure to propose mitigation strategies that will avoid significant adverse consequences to utility operation or assets.

Other Collaborative Cybersecurity Initiatives

There are a number of collaborative efforts involving private sector, academia, national laboratories, consumer groups, and others looking at cybersecurity and Smart Grid issues. Federal agencies are also working together to ensure communication and share resources on these issues, and helping to fund industry cybersecurity initiatives.[71] Two of the major DOE-led initiatives are discussed below.

DOE is home to the Federal Smart Grid Task Force (FSGTF) established under EISA to "ensure awareness, coordination, and integration of the diverse activities of DOE's Office of Electricity Delivery and Energy Reliability (OE) and elsewhere in the federal government related to Smart Grid technologies, practices, and services."[72] FERC, the Department of Commerce, the Environmental Protection Agency, the Department of Agriculture, and the Department of Defense are also members of the FSGTF.

DOE also funds (with OE leading DOE's participation) the "Trustworthy Cyber Infrastructure for the Power Grid" (TCIPG) initiative, working with DHS and industry.[73] TCIPG is led by the academic sector as a public-private research partnership, supporting the development of resilient and secure Smart Grid systems, leveraging research previously funded by the National Science Foundation:

> TCIPG's research plan is focused on securing the low-level devices, communications, and data systems that make up the power grid, to ensure trustworthy operation during normal conditions, cyber-attacks, and/or power emergencies. At the device level, new key functionality is being designed in hardware in order to detect attacks and failures and to restore proper system operation. Likewise, virtual machine technology is being developed and adapted for advanced power meters in order to permit new power use scenarios while preserving privacy. At the protocol level, new techniques are being developed to detect, react to, and recover from cyber attacks that occur while preserving integrity, availability, and real-time requirements.

Participating in TCIPG are the University of Illinois at Urbana-Champaign, Dartmouth College, Cornell University, the University of California at Davis, and Washington State University. Current TCIPG projects are focused on developing "trustworthy cyber infrastructure and technologies" for active demand management and wide-area monitoring and control, response to and managing cyber events, and risk/security assessments.

CURRENT LEGISLATION

Several bills introduced in the 112th Congress addressing the cybersecurity of the Smart Grid are summarized below.

Senate

The Cyber Security and American Cyber Competitiveness Act of 2011[74] (S. 21) would secure the United States against cyberattack and protect sensitive information and the identities of U.S. citizens and business. The bill focuses on the increasing dependence of the United States on information technology, and highlights a growing need for public awareness and preparedness of cyber-threats.

The Cybersecurity and Internet Freedom Act of 2011[75] (S. 413) is intended to enhance the "security and resiliency of the cyber and communications infrastructure" of the United States. Among its provisions, the bill would establish within DHS a National Center for Cybersecurity and Communications (NCCC) to protect the federal and national information infrastructure. The duties of the NCCC Director would include the identification and evaluation of cyber-risks to critical infrastructure (covered by the bill) on a continuous, sector-by-sector basis. The NCCC would also issue regulations establishing risk-based security performance requirements to secure covered critical infrastructure against cyber risks. Each owner or operator of covered critical infrastructure would be required to certify to the NCCC Director that approved security and emergency measures had been implemented. Civil penalties for non-compliance would be established by the bill. The President would be authorized by the bill to declare a national cyberemergency, and would be required to notify owners and operators of critical infrastructure about the nature of a cyber-threat if a national cyber-emergency was declared.

The Cyber Security Public Awareness Act of 2011[76] (S. 813) is intended to promote the awareness of cybersecurity. Among its provisions, the bill requires a report from be DHS on technical options to enhance the security of information networks of entities which own or manage critical infrastructure. The bill also requires a report from the Secretary of Homeland Security in consultation with the Secretary of Defense and the National Intelligence Director concerning the threat posed by a cyberattack to disrupt the national electrical grid and its implications to national security, the options for quick recovery to provide for national security as well as for restoration of all electrical service in the United States, and the development of a plan to prevent disruption of the U.S. grid by a cyberattack.

House

The National Infrastructure Development Bank Act of 2011[77] (H.R. 402) aims to facilitate infrastructure development. The bill identifies the nation's energy infrastructure as an area in need of modernization. Eligibility is extended to Smart Grid projects as energy infrastructure development.

The Secure High-voltage Infrastructure for Electricity from Lethal Damage Act[78] (H.R. 668) seeks to protect the bulk power system and electric infrastructure critical to the defense and wellbeing of the United States from EMP threats (such as nuclear weapons), or natural geomagnetic disturbance (GMD) vulnerabilities (such as solar flares). The bill proposes to authorize FERC to order the ERO to take emergency measures to protect the reliability of the bulk power system upon identification of an imminent threat to grid security from an EMP attack or GMD event. FERC would be authorized to allow utilities to recover "substantial costs" incurred from compliance with such an emergency order. The bill would also authorize FERC to issue rules or orders to protect against such grid security vulnerabilities not covered by ERO reliability standards (and would rescind the FERC rule or order when a sufficient ERO reliability standard was in place). FERC would be required to direct the ERO to propose reliability standards to protect against "reasonably foreseeable" EMP or GMD events. The ERO would also be required to propose reliability standards (balancing risks and costs) to address availability of large transformers capable of restoring reliable grid operations after an EMP or GMD event, providing entities (either individually or jointly) which own such transformers to ensure their adequate availability in case such transformers are destroyed or disabled by such an EMP or GMD event. The President would be required to identify up to 100 electric infrastructure facilities critical in supplying power for the defense of the United States which would be subject to FERC rules and orders related to EMP or GMD events. Recovery of full incremental costs incurred in compliance with such FERC rules and orders would be allowed for owners of the identified defense critical electric facilities.

White House Legislative Proposal on Cyberspace

The Obama administration has launched its own comprehensive proposals for legislation on cybersecurity.[79] Protection of critical infrastructure (including the electric grid, financial systems, and transportation networks) is among the main proposals. Other initiatives are suggested to better report cyber-breaches so that consumers are aware that their personal data may have been accessed by intruders, and "synchronized" penalties for cyber-crime with mandatory minimum punishments for cyber-intrusions into critical infrastructure.

The Administration wants better preparedness and coordination between federal agencies, state governments, and industry in order to prevent cyberattacks against critical infrastructure. The proposals include development of cybersecurity standards and enhanced information sharing about cyber threats. The Administration also proposes legislative provisions for DHS to help industry, states, and local government to recover from a cyber-intrusion and fix the resulting damage, or provide advice on building a better network. DHS would also review cyber-protection strategies and requirements, making public its findings on readiness (while not revealing the specific nature of any deficiencies).

Additional Comments

Much of the federal efforts related to the cybersecurity of the smarter grid relate to efforts to stay ahead of cyber-threats. But such is the complexity of the cyber-threat universe (with defined threats, perceived weaknesses, and the possibility of continual upgrades of both hardware and software systems) that staying ahead of the threats could quickly evolve into a rapidly growing, never-ending cost center for utilities. While a key to cost control is in assessment of the likely risks, cost recovery for reliability under current federal law is based on mandatory compliance with CIP cybersecurity standards for reliability purposes.

Cost recovery for Smart Grid expenditures or upgrades is voluntary under current law. Such discretionary treatment may possibly result in liability claims if damages result from an attack on Smart Grid systems. While the "Support Anti-Terrorism by Fostering Effective Technologies Act" of 2002[80] provides "risk management" and "litigation management" protection for qualified antiterrorism technologies and others in the supply and distribution chain, such protections likely do not apply to electric utilities as adopters of Smart Grid technologies and devices.

Current legislative and Administration proposals largely speak to a prioritization of critical infrastructure facilities, with consideration to how these relate to national security, and national economic recovery. However, recoverability from a cyberattack on the scale of something which might take down a significant portion of the grid will likely be very difficult. Maintaining a ready inventory of critical spare parts in close proximity to key installations would likely prove useful to quickening recovery efforts from some types of attack.

The current Smart Grid cybersecurity discussions largely focuses on the security of central station power plants and transmission systems. However, the future Smart Grid may increasingly depend on renewable energy, fuel cells, and other distributed resources like energy storage as these technologies are increasingly integrated into the nation's energy framework. The development of the Smart Grid with distributed and renewable power generation resources may add a level of security to the grid, since these resources do not have the fuel requirements of fossil generation. Damage to the fossil fuel delivery networks would likely impair operation of central station generating plants, depending on how much of an inventory of fuel is stored on-site. This greater diversity of resource options would likely further enhance the Smart Grid's expected improvement in reliability due a greater diversity of resource options, joining together these newer elements with traditional power stations in the power grid of the future. But the characteristics that these elements bring to the system could be considered in the design of CIP standards and protocols.

Although this report has focused on cybersecurity aspects of a smarter electrical grid, technologies being developed for use by the Smart Grid could also be used by natural gas pipeline, water supply, and telecommunications systems. Consideration could be given to applying similar control system device and system safeguards to these other vital utility systems. Consideration could be given to applying similar control system device and system safeguards to these other vital utility systems. The electricity grid is connected to (and largely dependent on) these other resource and communications providers as the vast majority of power production today needs water for steam generation or natural gas for fuel. Similarly, natural gas pipelines use many electrically powered compressors to maintain pipeline pressure and move gas through the system, and the nation's water systems is a notable consumer of electricity for distribution and supply purposes. These interdependent systems are also vulnerable to natural hazards such as earthquakes and tornadoes. The remote sensing and other technologies expected to be employed by the Smart Grid could also help mitigate the effects of regional natural disasters on critical infrastructure, as these are expected to be capable of detecting outages and optimizing power flows to minimize damage and enable a more complete recovery.

End Notes

[1] EIA, Glossary, http://www.eia.gov/tools/glossary/index.cfm?id=B.

[2] EIA, Glossary, http://www.eia.gov/tools/glossary/index.cfm?id=B.

[3] As of 2007, there were 210 investor-owned electric utilities, 2,009 publicly-owned electric utilities, 883 consumerowned rural electric cooperatives, and nine federal electric utilities. Energy Information Administration (EIA), Electric Power Industry Overview 2007, http://www.eia.doe.gov/electricity/page/prim2/toc2.html.

[4] The bulk power system makes it possible for utilities to engage in wholesale (sales for resale) electric power trade. U.S. Energy Information Administration, Overview—Power Transactions & Interconnected Networks, 2011, http://www.eia.gov/cneaf/electricity/page/ prim2/chapter7.html.

[5] Massachusetts Institute of Technology, Retrofitting of Coal-Fired Power Plants for CO2 Emissions Reductions, March 23, 2009, http://web.mit.edu/mitei/docs/reports/meeting-report.pdf.

[6] Thomas A. Prevost and David J. Woodcock, Transformer Fleet Health and Risk Assessment, Weidman Electrical Technology, IEEE PES Transformers Committee Tutorial, March 13, 2007, http://grouper.ieee.org/groups/transformers/info/S07/S07-TR_LifeExtension.pdf.

[7] K. Anderson, D. Furey, and K. Omar, Frayed Wires: U.S. Transmission System Shows its Age, Fitch Ratings, October 25, 2006.

[8] Key legislation include the Public Utility Regulatory Policies Act of 1978, the Energy Policy Acts of 1992 and 2005, and the Energy Independence and Security Act of 2007.

[9] See Title XIII at 42 U.S.C. 17381 *et seq.*

[10] CRS Report R41674, *Terrorist Use of the Internet: Information Operations in Cyberspace*, by Catherine A. Theohary and John Rollins.

[11] United States Department of Energy, Office of Electric Transmission and Distribution, *"Grid 2030" A National Vision for Electricity's Second 100 Years*, July 2003, p. 27, http://www.oe.energy.gov/DocumentsandMedia/Electric_Vision_Document.pdf.

[12] Pacific Northwest National Laboratory, *North American SynchroPhasor Initiative*, May 2011, http://www.naspi.org/.

[13] U.S. Department of Energy, *Strategic Plan*, May 2011, http://www.energy.gov /news/documents/ DOE_ Strategic Plan.pdf.

[14] "Demand Response" allows retail customers to participate in electricity markets by giving them the ability to respond to prices as they change over time—either daily or hourly in most instances. See http://www.oe.energy.gov/demand.htm.

[15] The Eastern, Western and Texas interconnections of the U.S. grid. See http://www.eia.doe.gov/cneaf/electricity/chg_stru_update/fig7.html.

[16] Critical Foundations: Protecting America's Infrastructures, The President's Commission on Critical Infrastructure Protection , October 1997, http://www.fas.org/sgp/library/pccip.pdf. (CF)

[17] A system of remote control and telemetry used to monitor and control the transmission system. See http://www.eia.doe.gov/tools/glossary/index.cfm?id=S.

[18] CF, p. 12.

[19] The Smart Grid Interoperability Panel Cyber Security Working Group, Introduction to NISTIR 7628, National Institute of Standards and Technology, Guidelines for Smart Grid Cyber Security, September 2010, http://csrc.nist.gov/publications/nistir/ir7628/introduction-to-nistir-7628.pdf. (NISTIR)

[20] NISTIR, p. 9.

[21] Siobhan Gordon, Electricity Grid in U.S. Penetrated by Spies, The Wall Street Journal, April 8, 2009, http://online.wsj.com/article/SB123914805204099085.html.

[22] Ibid.

[23] "[A worm is a]n independent computer program that reproduces by copying itself from one system to another across a network. Unlike computer viruses, worms do not require human involvement to propagate." See http://www.gao.gov/new.items/d09661t.pdf, p.6.

[24] "A successful attack by a software application such as the Stuxnet worm could result in manipulation of control system code to the point of inoperability or long-term damage. Should such an incident occur, recovery from the damage to the computer systems programmed to monitor and manage a facility and the physical equipment producing goods or services could be significantly delayed. Depending on the severity of the attack, the interconnected nature of the affected critical infrastructure facilities, and government preparation and response plans, entities and individuals relying on these facilities could be without life sustaining or comforting services for a long period of time. The resulting damage to the nation's critical infrastructure could threaten many aspects of life, including the government's ability to safeguard national security interests." CRS Report R41524, The Stuxnet Computer Worm: Harbinger of an Emerging Warfare Capability, by Paul K. Kerr, John Rollins, and Catherine A. Theohary.

[25] "The Stuxnet trojan/virus is the first publicly known 'worm' to target industrial control systems. The threat posed by Stuxnet has been portrayed as beyond anything seen before. Its goal was to sabotage a real-world industrial plant, not disrupt abstract IT systems. It was aimed at industrial control systems with the intention to reprogram PLCs [programmable logic controllers—devices used for control of processes] in a manner that would sabotage the plant, hiding the changes from programmers or users. Stuxnet has highlighted the potential to directly attack industrial control systems used in critical national infrastructure, including energy, water, and transport sectors." Richard Piggin, Control Network Security Lessons from Stuxnet, Consulting-Specifying Engineer, February 3, 2011, http://m.csemag.com/index.php?id=2832&tx_ttnews%5Btt_news%5D= 42715&cHash=42da992b88.

[26] Military Photos.net, AURORA Test Validated Fears of Dept. of Homeland Security, October 1, 2007, http://www.militaryphotos.net/forums/showthread.php?121081-AURORA-test-validated-fears-of-Dept.-of-Homeland-Security.

[27] U.S. Government Accountability Office, INFORMATION SECURITY Cyber Threats and Vulnerabilities Place Federal Systems at Risk, GAO-09-661T, May 5, 2009, p. 12, http://www.gao.gov/new.items/d09661t.pdf.

[28] Jeffrey Carr, "Ember Needs a Wake-Up Call from the CIA," Forbes, January 31, 2011, http://blogs.forbes.com /jeffreycarr/2011/01/31/ember-needs-a-wake-up-call-from-the-cia/.

[29] A method of attack that denies system access to legitimate users without actually having to compromise the targeted system. From a single source, the attack overwhelms the target computers with messages and blocks legitimate traffic. It can prevent one system from being able to exchange data with other systems or prevent the system from using the Internet. See http://www.gao.gov/new.items/d09661t.pdf, p.5.

[30] A high-tech scam that frequently uses spam or pop-up messages to deceive people into disclosing sensitive information. Internet scammers use e-mail bait to "phish" for passwords and financial information from the sea of internet users. See http://www.gao.gov/new.items/d09661t.pdf, p. 6.

[31] NISTIR. Page 7.

[32] James St. Pierre, *Smart Grid Interoperability Panel Cyber Security Working Group*, National Institute of Standards and Technology, Presentation at National Electricity Forum, Washington, DC, February 16-17, 2011, p. 10, http://www.nationalelectricityforum. org/pdfs/NationalElectricityForum_StPierre_2011Feb08.pdf.

[33] Edison Electric Institute, *Spare Transformers*, 2011, http://www.eei.org/ourissues/ ElectricityTransmission /Pages/SpareTransformers.aspx.

[34] Software that is embedded in a hardware device that allows reading and executing the software, but does not allow modification, e.g., writing or deleting data by an end user. See http://www.its.bldrdoc.gov/fs-1037/dir-015/_2236.htm.

[35] Committee on the Societal and Economic Impacts of Severe Space Weather Events, National Research Council, *Severe Space Weather Events—Understanding Societal and Economic Impacts: A Workshop Report*, National Academies, May 22-23, 2008, http://books.nap.edu/catalog.php?record_id=12507.

[36] Ibid.

[37] Yousaf M. Butt, "The EMP Threat: Fact, Fiction, and Response," *Space Review*, February 1, 2010, http://www.thespacereview.com/article/1553/2.

[38] 18 U.S.C. § 1030.

[39] Interoperability can be defined as the capability of two or more networks, systems, devices, applications, or components to share and readily use information securely and effectively with little or no inconvenience to the user. GridWise Architecture Council, *Interoperability Path Forward Whitepaper*, November 30, 2005, http://www.gridwiseac.org/pdfs/ interoperability_path_whitepaper_v1_0.pdf.

[40] The electrical generation resources, transmission lines, interconnections with neighboring systems, and associated equipment, generally operated at voltages of 100 kV or higher. http://www.nerc.com/files/Glossary_ 2009April20.pdf.

[41] North American Electric Reliability Corporation, Rules of Procedure, March 21, 2008, http://www.nerc.com/files/NERC_Rules_of_Procedure_EFFECTIVE_20080321.pdf.

[42] The maximum civil penalty which can be assessed for violation of reliability standards is $1 million per day. Revised Policy Statement on Penalty Guidelines, 132 FERC ¶ 61,216, issued September 17, 2010.

[43] North American Electric Reliability Corporation, Reliability Standards for the Bulk Electric Systems of North America, April 20, 2010, http://www.nerc.com/files/Reliability_Standards _Complete_Set.pdf.

[44] Programmable electronic devices and communication networks including hardware, software, and data. http://www.nerc.com/files/Glossary_2009April20.pdf.

[45] In FERC Order No. 743, the North American Electric Reliability Corporation was directed to revise its definition of the term "bulk electric system" to ensure that the definition encompasses all facilities necessary for operating an interconnected electric transmission network. The revision will be made in accordance with NERC's Reliability Standards development process. See http://www.ferc.gov/whats-new/comm-meet/2010 /111810/E-2.pdf.

[46] The standards produced by the International Electrotechnical Commission (IEC) focus on the informational models and procedures necessary to promote the reliable and efficient operation of the Smart Grid. Standard IEC 62351 specifically addresses the cyber security of communications protocols.

[47] Federal Energy Regulatory Commission, Smart Grid Interoperability Standards—Supplemental Notice Requesting Comments, February 16, 2011, http://www.ferc.gov/ EventCalendar/Files/20110228084004-supplemental-notice.pdf.

[48] U.S. Government Accountability Office, Electricity Grid Modernization: Progress Being Made on Cybersecurity Guidelines, but Key Challenges Remain to be Addressed, GAO-11-117, January 12, 2011, http://www.gao.gov/products/GAO-11-117.

[49] Ibid, p. 27.

[50] Federal Energy Regulatory Commission, Letter from FERC Chairman Wellinghoff to David A. Powner, Director, Information Technology Management Issues for the United States Government Accountability Office, December 23, 2010, http://www.ferc.gov/industries/ electric/indus-act/smart-grid/wellinghoff-response-GAO.pdf.

[51] Letter from G. Arnold, NIST to FERC Chairman J. Wellinghoff, dated April 7, 2011. See http://www.nist.gov /smartgrid/upload/FERC_Letter1.pdf.

[52] U.S. Department of Energy, Office of Inspector General, Audit Report, Federal Energy Regulatory Commission's Monitoring of Power Grid Cyber Security, January 2011, http://www.ig.energy.gov/documents/IG-0846.pdf. (DOEIG).

[53] DOEIG, pages 11-12.

[54] DOEIG, pages 21-22.

[55] Defined by NERC as "facilities, systems, and equipment which, if destroyed, degraded, or otherwise rendered unavailable, would affect the reliability or operability of the bulk electric system." See http://www.nerc.com/ files/Glossary_of_Terms_2011Mar15.pdf.

[56] NERC works with eight regional entities to improve the reliability of the bulk power system. The members of the regional entities come from all segments of the electric industry: investor-owned utilities; federal power agencies; rural electric cooperatives; state, municipal and provincial utilities; independent power producers; power marketers; and endues customers. These entities account for virtually all the electricity supplied in the United States, Canada, and a portion of Baja California Norte, Mexico. See http://www.nerc.com/page.php?cid=1%7C9%7C119.

[57] U.S. Department of Energy Office of Electricity Delivery and Energy Reliability, Study of Security Attributes of Smart Grid Systems—Current Cyber Security Issues, INL/EXT-09-15500, April 2009, http://www.inl.gov/scada/publications/d/securing_the_smart_grid_ current_issues.pdf.

[58] National Institute of Standards and Technology, NIST Framework and Roadmap for Smart Grid Interoperability Standards, Release 1.0, NIST Special Publication 1108, January 2010, http://www.nist.gov/public_affairs/releases/upload/smartgrid_interoperability_final.pdf.

[59] On July 16, 2009, the Commission issued a Smart Grid Policy Statement in which, among other things, the Commission explained its view that EISA does not make any Smart Grid standards mandatory and does not give the Commission authority to make or enforce any such standards. The Commission clarified that, under current law, its authority, if any, to make Smart Grid standards mandatory [or allow recovery in rates] derives from the Federal Power Act. Smart Grid Policy, 128 FERC ¶ 61,060, issued July 16, 2009. See http://www.ferc.gov/whats-new/comm-meet/2009/071609/E-3.pdf. (FERCSGP)

[60] The Commission also proposed to permit applicants to seek recovery of the otherwise stranded costs of legacy systems that are to be replaced by smart grid equipment. The Commission stated that an appropriate plan for the staged deployment of smart grid equipment, which could include appropriate upgrades to legacy systems where technically feasible and cost-effective, could help minimize the stranding of unamortized costs of legacy systems. The Commission therefore proposed that any request to recover stranded legacy system costs must demonstrate that such a migration plan has been developed. FERCSGP. Page 82.

[61] Department of Homeland Security, Safeguard and Secure Cyberspace, http://www.dhs.gov/ xabout/gc_1240609042614.shtm.

[62] Federal Energy Regulatory Commission, FY 2010 Performance and Accountability Report, November 2010, http://www.ferc.gov/about/strat-docs/2010-audit.pdf.

[63] See footnote 58.

[64] FERCSGP, p. 27.

[65] M. D. Garratt, Distributech Highlights Challenge of SmartGrid: What's in it for Consumers?, CleanMakesGreen.com, February 7, 2011, http://cleanmakesgreen.com/2011/02/07/ distributech-highlights-challengeof-smartgrid-what%E2%80%99s-in-it-for-consumers/.

[66] Business Wire, Smart Grid Cyber Security Market to Reach $3.7 Billion by 2015, According to Pike Research, Smart Grid Cyber Security, June 23, 2010, http://www.businesswire.com /portal/site/home/permalink/?ndmViewId=news_view&newsId=20100623005613&newsLang=en.

[67] Mathew J. Schwartz, California Proposes Smart Grid Data Privacy Standards, Information Week, May 18, 2011, http://mobile.informationweek.com/10247/show/ 8d96dc971 db09af9557e4ad59b4708df&t=f6de813a074007af2879af82d46659ef.

[68] Sherry Lichtenberg, Smart Grid Data: Must There Be Conflict Between Energy Management and Consumer Privacy?, National Regulatory Research Institute, December 2010, http://www.nrri.org/pubs/telecommunications/NRRI_smart_grid_privacy_dec10-17.pdf.

[69] Patricia Hoffman, Statement Before the U.S. Senate Committee on Energy and Natural Resources, Office of Electricity Delivery and Energy Reliability - U.S. Department of Energy, Cybersecurity of the bulk-power system and electric infrastructure, May 5, 2011, p. 5, http://energy.senate.gov/public/_ files/HoffmanTestimony050511.pdf.

[70] See http://www.eei.org/ourissues/ElectricityTransmission/Documents/cyber_security_ principles.pdf.

[71] Electric Light & Power.com, DOE Picks EPRI Collaborative to Lead Cybersecurity Project, December 27, 2010, http://www.elp.com/index/display/article-display/6433556709/articles/electric-light-power/policy-and_regulation/2010/09/DOE_picks_EPRI_collaborative _to_lead_cybersecurity_project_.html.

[72] U.S. Department of Energy, Federal Smart Grid Task Force, http://www.oe. energy.gov/ smartgrid_taskforce.htm.

[73] Trustworthy Cyber Infrastructure for the Power Grid, About TCIPG, 2011, http://tcipg.org/about-tcipg-trustworthycyber-infrastructure-power-grid.

[74] Introduced in January 2011 by Sen. Reid et al.

[75] Introduced in February 2011 by Sen. Lieberman et al.

[76] Introduced in April 2011 by Senators Whitehouse and Kyl.

[77] Introduced in January 2011 by Rep. DeLauro et al.

[78] Introduced in February 2011 by Rep. Franks et al.

[79] White House, Office of the Press Secretary, *Fact Sheet: Cybersecurity Legislative Proposal*, May 12, 2011, http://www.whitehouse.gov/the-press-office/2011/05/12/fact-sheet-cybersecurity-legislative-proposal.

[80] P.L. 107-296.

In: The Transformation of Electrical Power: Key Issues
Editor: Dawson B. Munoz
ISBN: 978-1-62618-619-4

Chapter 5

WEATHER-RELATED POWER OUTAGES AND ELECTRIC SYSTEM RESILIENCY*

Richard J. Campbell

SUMMARY

High winds, especially when combined with precipitation from seasonal storms, can cause damage to electricity utility systems, resulting in service interruptions to large numbers of electricity customers. While most such power outages are caused by damage from trees and tree limbs falling on local electricity distribution lines and poles, major power outages tend to be caused by damage to electricity transmission lines which carry bulk power long distances. Depending on the severity of the storm and resulting impairment, power outages can last a few hours or extend to periods of several days, and have real economic effects. Power outages can impact businesses (primarily through lost orders and damage to perishable goods and inventories), and manufacturers (mainly through downtime and lost production, or equipment damage). Data from various studies lead to cost estimates from storm-related outages to the U.S. economy at between $20 billion and $55 billion annually. Data also suggest the trend of outages from weather-related events is increasing.

Suggested solutions for reducing impacts from weather-related outages include improved tree-trimming schedules to keep rights-of-way clear, placing distribution and some transmission lines underground, implementing Smart Grid improvements to enhance power system operations and control, inclusion of more distributed generation, and changing utility maintenance practices and metrics to focus on power system reliability. However, most of these potential solutions come with high costs which must be balanced against the perceived benefits.

A number of options exist for Congress to consider which could help reduce storm-related outages. These range from improving the quality of data on storm-related outages, to a greater strategic investment in the U.S. electricity grid. Congress could empower a federal

* This is an edited, reformatted and augmented version of a Congressional Research Service publication, CRS Report for Congress R42696, from www.crs.gov, prepared for Members and Committees of Congress, dated August 28, 2012.

agency to develop standards for the consistent reporting of power outage data. While responsibility for the reliability of the bulk electric system is under the Federal Energy Regulatory Commission (as per the Energy Policy Act of 2005), no central responsibility exists for the reliability of distribution systems. One possible option could be to bring distribution systems under the Electric Reliability Organization for reliability purposes. Recovery after storm-related outages might be enhanced by a federal role in formalizing the review or coordination of electric utility mutual assistance agreements (MAAs). This would not necessarily mean federal approval of MAAs, but may help in the cooperative coordination of additional federal and state resources, especially in a wide, multi-state weather event. While there has been much discussion of transmission system inadequacies and inefficiencies, many distribution systems are in dire need of upgrades or repairs.

The cost of upgrading the U.S. grid to meet future uses is expected to be high, with the American Society of Civil Engineers estimating a need of $673 billion by 2020. While the federal government recently made funding available of almost $16 billion for specific Smart Grid projects and new transmission lines under the American Recovery and Reinvestment Act of 2009, there has not been a comprehensive effort to study the needs, set goals, and provide targeted funding for modernization of the U.S. grid as part of a long-term national energy strategy. Such an effort would also require decisions about the appropriate roles of government and the private sector.

Power delivery systems are most vulnerable to storms and extreme weather events. Improving the overall condition and efficiency of the power delivery system can only serve to improve the resiliency of the system, and help hasten recovery from weather-related outages.

Ultimately, however, electric utilities are responsible for this infrastructure. They are in the business of selling electricity, and they cannot sell electricity if their power delivery systems are out of service.

INTRODUCTION

High winds, especially when combined with precipitation from seasonal storms, can cause damage to electricity utility systems, resulting in service interruptions to large numbers of electricity customers. While most such power outages[1] are caused by damage from trees and tree limbs falling on local electricity distribution lines and poles, major power outages tend to be caused by damage to electricity transmission lines, which carry bulk power long distances.

Depending on the severity of the storm and resulting impairment, power outages can last a few hours or extend to periods of several days. This in turn can have real economic effects, as power outages can impact businesses (primarily through lost orders and damage to perishable goods and inventories), and manufacturers (mainly through downtime and lost production, or equipment damage).

Potential issues for Congress concern the resiliency of the electric power system, and whether the economic impacts of storm-related power outages require additional measures to be undertaken by the federal government in mitigating storm-related outages. Congress has already recognized the importance of the reliable operation of the bulk power system with the

Energy Policy Act of 2005[2] (EPACT) authorization of an Electric Reliability Organization[3] (ERO).

This report will focus on the impacts of sustained power outages as might result from the result of seasonal storms, and whether there is a role for the federal government in hastening the restoration of power from weather-related outages.

ANATOMY OF WEATHER-RELATED POWER OUTAGES

Electric power is generated and sent over transmission lines to substations which reduce the voltage levels for distribution to end-use customers, as shown in Figure 1. The network of cables enabling electric power to be sent to customers generally exists in an exterior or "above ground" environment largely exposed to the elements. As such, power outages can result from seasonal storms which often combine the furies of wind, rain, snow, or ice. The more severe weather events usually cause the greatest damage to electric power transmission and distribution infrastructure as damage can result from trees or branches falling on electricity lines. While data on storm-related power outages exist, they are not generally considered to be complete or well-characterized with regard to the cause of the outage event. It has been estimated that 90% of customer outage-minutes are due to events which affect local distribution systems.4 However, the remaining 10% stem from generation and transmission problems, which can cause wider-scale outages affecting larger numbers of customers.

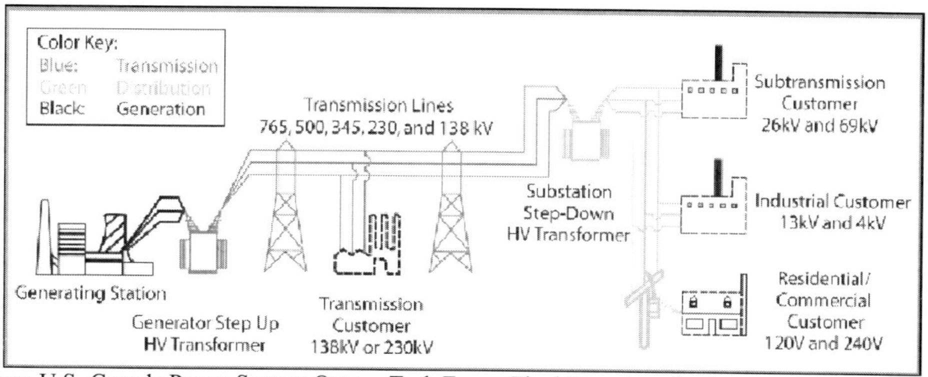

Source: U.S.-Canada Power System Outage Task Force, Final Report on the August 14, 2003, Blackout in the United States and Canada: Causes and Recommendations, April 2004, p. 5, https://reports.energy.gov/BlackoutFinalWeb.pdf.

Figure 1. Electric Power System Elements.

U.S. Storm-Related Outages

The North American Electric Reliability Corporation (NERC) requires electric utilities to report events which cause disturbances that interrupt service (i.e., power outages) of more than 300 MegaWatts (MW) or affect 50,000 customers or more. A University of Vermont analysis5 of NERC data describes 933 events causing outages from the years 1984 to 2006, and is presented in Table 1.

According to the Vermont study, almost 44% of the events in the period were weather-related (i.e., caused by tornado, hurricane/tropical storm, ice storm, lightning, wind/rain, or other cold weather). The study noted that the data include many events smaller than the NERC reporting threshold. It also noted that some of the reported events have "multiple initiating" causes, since some events (such as lightning) can trigger other outages or operator errors.

Table 1. Large Blackouts in the United States

Statistics for Outage Cause Categories

	% of events	Mean size in MW	Mean size in customers
Earthquake	0.8	1,408	375,900
Tornado	2.8	367	115,439
Hurricane/Tropical Storm	4.2	1,309	782,695
Ice Storm	5	1,152	343,448
Lightning	11.3	270	70,944
Wind/Rain	14.8	793	185,199
Other cold weather	5.5	542	150,255
Fire	5.2	431	111,244
Intentional attack	1.6	340	24,572
Supply shortage	5.3	341	138,957
Other external cause	4.8	710	246,071
Equipment Failure	29.7	379	57,140
Operator Error	10.1	489	105,322
Voltage reduction	7.7	153	212,900
Volunteer reduction	5.9	190	134,543

Source: Trends in the History of Large Blackouts in the United States, http://www. uvm.edu/~phines/ publications/ 2008/Hines_2008_blackouts.pdf.

Notes: Totals are greater than 100% because some events fall into multiple initiating-event categories.

A 2004 study[6] by Lawrence Berkeley Laboratory (LBL) looking at power interruptions characterized power outages as being of *short* duration lasting less than five minutes, and *sustained* duration outages[7] lasting longer than five minutes (and extending to hours or days). Power outages caused by storm-related events can vary in duration but tend to be sustained disruptions. The study noted that weather-related events are not always captured in power outage data.[8]

The U.S. Department of Energy (DOE) maintains its own database of grid disturbance events.[9] A recent analysis[10] by LBL's Evan Mills of the DOE database shows an increasing number of outages from 1992 to 2010, all dominated by weather-related events (see *Figure* 2). According to Mills, approximately 78% of the reported 1,333 electric grid disruptions in the period were weather-related (i.e., caused by temperature extremes, ice/snow/winter storm, thunderstorm/tornado/lightning, windstorm/hurricane/severe storm, and undefined weather events), and the grid disruptions affected 178 million metered customers.

According to graphs shown in *Figure 2,* the number of observed outages from weather-related incidents seems to be increasing. Mills believes the reasons for the increased trend in outages may be due to a combination of power grid deterioration and a real increase in the number of observed extreme weather events.[11] According to graphs shown in *Figure 2*, the number of observed outages from weather-related incidents seems to be increasing. Mills believes the reasons for the increased trend in outages may be due to a combination of power grid deterioration and a real increase in the number of observed extreme weather events.[11]

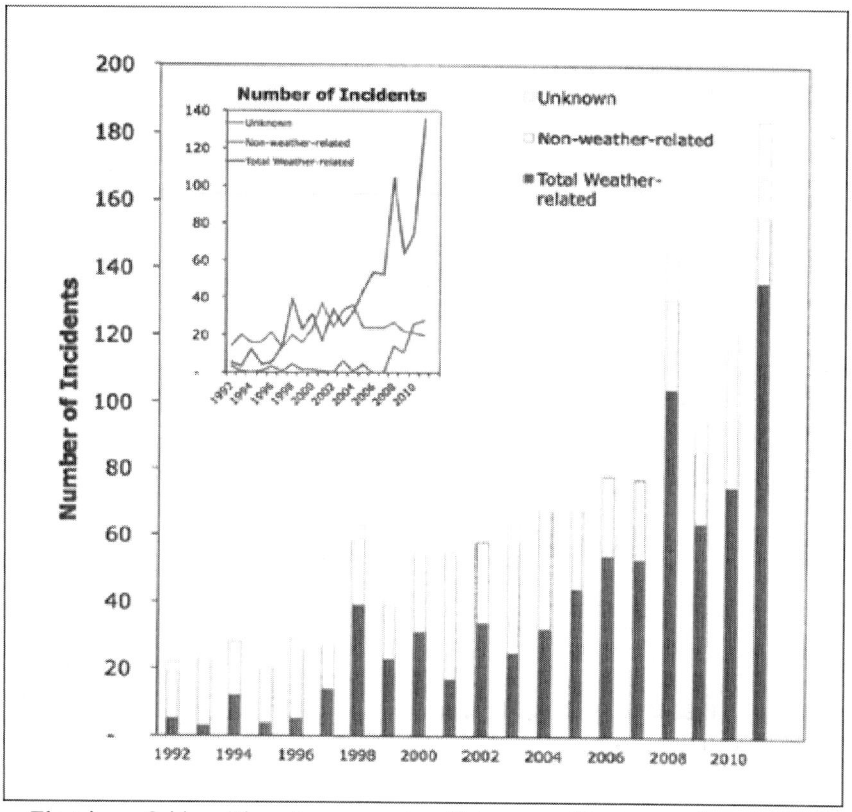

Source: Electric Grid Disruptions and Extreme Weather. See http://evanmills.lbl.gov/presentations/Mills-GridDisruptions-NCDC-3May2012.pdf.

Notes: Historical "Grid Disturbance" data from the U.S. Department of Energy, Energy Information Administration. Form OE-417, "Electric Emergency Incident and Disturbance Report" (and before 1978 from the National Electric Reliability Council, Disturbance Analysis Working Group).

Figure 2. Significant U.S Grid Weather-Related Grid Disturbances. (With Inset of Non-Weather- vs. Weather-Related Outage Comparison).

Outage Rates in Other Countries

The United States is generally considered to have one of the industrial world's most reliable electric power systems. However, when compared statistically to other nations, the U.S. grid does not necessarily meet those expectations.

There are two main indices generally used to measure reliability. The *system average interruption duration index*[12] (SAIDI) represents the average amount of time per year that power supply to a customer is interrupted, expressed in minutes per customer per year. The *system average interruption frequency index*[13] (SAIFI) represents the average number of times per year that the supply to a customer is interrupted, expressed as interruptions per customer per year. However, there is a lack of consistency in how the inputs to these indices are measured, both domestically and internationally. Much of the discrepancy again concerns whether or how storm-related outage events are counted as outage events. Some jurisdictions, both in the United States and internationally, consider storm-related outages as "extreme" events, and thus are not included in power outage statistics. Additionally, what is considered as unusual weather in one region may not be counted as unusual in another region.

> It should be noted, however, that weather circumstances that occur occasionally should not be considered as exceptional events. For example, snowstorms are not an exceptional event in Sweden, but could be seen as an exceptional event in southern Greece. Similarly, very hot temperature for sustained periods of time is not an exceptional event in Greece, but could be considered so in Sweden. Lightning should not be treated as an exceptional event anywhere in Europe.[14]

SAIDI and SAIFI reliability indices for nine industrial countries are summarized in Table 2. As can be seen, the United States has the highest average annual outage time per customer, and the third-highest average annual number of supply outages per customer.

It should be noted that many European countries consider the threshold for defining an extended outage as longer than three minutes[15] instead of the five minute standard generally used in the United States. European countries also commonly have a single regulatory body which is responsible for reliability of the entire nation's power system.

These authorities generally authorize expenditures for improving the reliability of service.

Table 2. Comparison of International Reliability Indices

Country	SAIDI	SAIFI
United States	240	1.5
Austria	72	0.9
Denmark	24	0.5
France	62	1.0
Germany	23	0.5
Italy	58	2.2
Netherlands	33	0.3
Spain	104	2.2
United Kingdom	90	0.8

Source: Galvin Electricity Initiative, Electric Reliability: Problems, Progress and Policy Solutions. See http://www.galvinpower.org/sites/default/files/Electricity_ Reliability_031611.pdf.

Overseeing the reliability of the bulk power system in the United States is the responsibility of the Federal Energy Regulatory Commission (FERC) as per EPACT, but responsibility for maintaining the reliability of the nation's electricity distribution systems is the responsibility of many state and local regulatory bodies. It is also likely that the physical size of the United States, and the historical weather extremes possible in any given season across the country, contribute to the U.S. standing in the statistics of *Table 2.*

However, the generally degraded condition of the U.S. electric grid's infrastructure also likely contributes to the statistical quandary. The American Society of Civil Engineers (ASCE) has labeled the nation's grid as a "patchwork system" that may ultimately break down without a massive investment of an estimated $673 billion by 2020.[16] ASCE estimated that over the next decade, there is a gap between what it viewed as needed and current spending annually of $12.3 billion for power generation, $37.3 billion for transmission systems, and $57.4 billion for distribution systems.[17]

ECONOMIC COSTS ASSOCIATED WITH STORM-RELATED POWER OUTAGES

Power outages can impact electricity consumers primarily through property loss and business disruption. This can result in lost orders, and damage to perishable goods and inventories for businesses. Power outages can critically affect manufacturing operations mainly through downtime as workers are idled, and potentially damage equipment and production processes. The loss of electricity can also disrupt many other electric power-dependent activities sometimes taken for granted (such as traffic control) which can result in further economic impacts.

A study in 2001 by the Electric Power Research Institute (EPRI) estimated the costs to the economy of power disturbances.[18] The study found that three sectors of the economy were particularly vulnerable to power outages:

> The digital economy (DE). This sector includes firms that rely heavily on data storage and retrieval, data processing, or research and development operations. Specific industries include telecommunications, data storage and retrieval services (including collocation facilities or Internet hotels), biotechnology, electronics manufacturing, and the financial industry.
>
> Continuous process manufacturing (CPM). This sector includes manufacturing facilities that continuously feed raw materials, often at high temperatures, through an industrial process. Specific industries include paper; chemicals; petroleum; rubber and plastic; stone, clay, and glass; and primary metals.
>
> Fabrication and essential services (FES). This sector includes all other manufacturing industries, plus utilities and transportation facilities such as railroads and mass transit, water and wastewater treatment, and gas utilities and pipelines.

These three sectors accounted for almost 40% of U.S. gross domestic product[19] in 2001, but disruptions to especially the DE and FES sectors impact other sectors of the economy through the services they provide.

EPRI recognized that the economic cost of power outages was largely related to the length of the outage (while noting even short duration outages of a few minutes could have large costs), and estimated the average cost of a one-hour outage for manufacturing and DE firms at $7,795 per firm. While the estimate recognizes that typically most companies experienced a much smaller loss (with a one-hour outage costing 56% of all DE, CPM, and FES establishments less than $500), the number was weighted by the much larger losses experienced by a few companies. An estimated 5% of establishments in these sectors incurred costs from a one-hour outage of $20,000 or more, with costs for individual establishments ranging as high as $1.5 million.[20] EPRI estimated the annual cost of outages across all U.S. business sectors at $104 billion to $164 billion in 2001. With EPRI estimating the average number of power outages in a year over five minutes at 43% (see *Figure 3*), a reasonable estimate for the annual economic cost yields a possible $20 billion to $31 billion (using the Hines estimate of 44% of outages as weather-related) as a lower range. On the higher end (using Mills's estimate of 78% of outages as weather-related), an estimate of the cost yields $35 billion to $55 billion for annual weather-related outage costs.

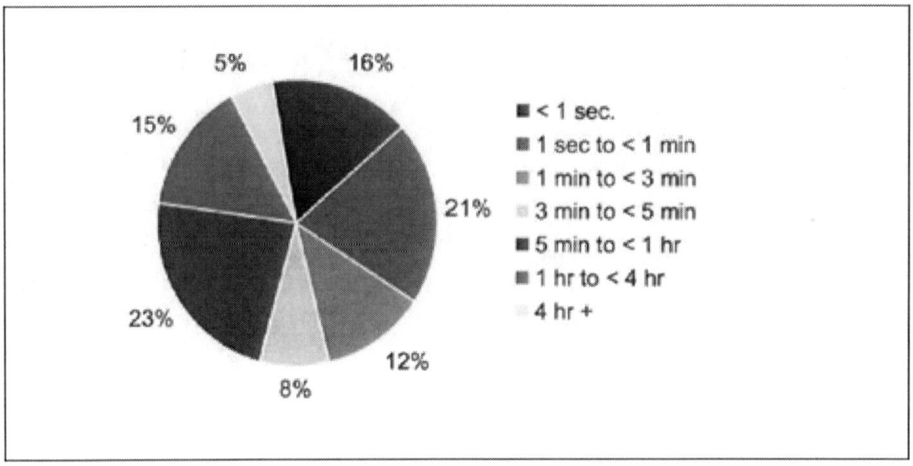

Source: The Cost of Power Disturbances to Industrial & Digital Economy Companies. See http://www.onpower.com/pdf/EPRICostOfPowerProblems.pdf.

Figure 3. Average Length of Power Outages in a Year.

The 2004 LBL report estimated (in its base case) economic costs of sustained power interruptions at $26 billion, with the outages resulting from all causes including weather. In recognition of uncertainties in the data, LBL stated that the true cost of sustained outages could be higher or lower by tens of billions of dollars.[21]

Reducing Storm-Related Outages

Outages are largely a result of damage to distribution systems, which are generally exposed to the elements. Storms can inflict damage to electric power delivery systems in a variety of ways. Wind and rain from a hurricane-type event tends to cause different types of power system delivery failures than snow and ice. This section will look at various scenarios for reducing storm-related outages.

Tree-Trimming Schedules

Many storm-related outages are a result of trees contacting or damaging power delivery systems. This is a common mode of electric service disruption especially affecting distribution systems (see *Figure 1*). To minimize the chances of power line contacts with trees, utilities typically trim tree branches to maintain a right-of-way[22] free of impinging vegetation.

To maintain reliability of the bulk power system and reduce the risk of outages, NERC has a vegetation management standard which applies to high-voltage transmission lines of 200 kiloVolts (kV) or higher, requiring the transmission line owner to have a vegetation management plan.[23] *Minimum* vegetation clearance distances are calculated by formula, varying from 0.82 feet (for a 69 kV line) to 8.06 feet (for a 765 kV line).[24] NERC's tree-trimming standard does not apply to low-voltage distribution lines, which are typically less than 69 kV.

NERC's reliability rules do not apply to distribution utilities which are commonly subject to state jurisdiction.Consequently, there is no national law mandating minimum standards for vegetation management around distribution lines. Schedules for tree-trimming have come under scrutiny as some utilities appear to be increasing the period between these operations as they seek to reduce maintenance costs.

Some state and local jurisdictions have instituted laws or regulations for tree-trimming or vegetation management. For example, New York and several states in New England require electric distribution utilities to submit a plan for tree-trimming and vegetation management for approval by a state utility commission.

> Although each company develops its own plan, subject to state approval, they tend to maintain an eight- to 10-foot clearance around the sides and bottom of a power line and a 12- to 15-foot clearance above it. Trimming cycles range from every four to seven years, with additional trimming and vegetation management for areas that have been problematic in the past.25

Nevertheless, vigilant maintenance of vegetation clearances for power line rights-of-way are generally seen as a good practice that may help to minimize storm-related outages.

Undergrounding of Distribution/Transmission Lines

The practice of putting electricity distribution lines underground has come to be called "undergrounding." In both urban and suburban areas, the practice is common for new construction or in areas being redeveloped to improve aesthetics. Undergrounding is sometimes considered as an option to improve the reliability of service in some areas believed to be susceptible to storm damage.

Underground transmission can be used in urban areas or in other locations where the use of overhead transmission lines may be technically challenging or undesirable. Examples of these situations may include instances when traversing an overland route is difficult, where rivers or natural obstacles exist, or when the line passes through land with a scenic or environmental value judged too great to disturb with an overhead transmission line. Underground transmission may also be less of a regulatory obstacle for these reasons. Underground lines are also safer for wildlife (i.e., a lessened risk of electrical shock) and have far fewer long-term visual impacts.

The cost of burying power lines has been estimated as being from ten to twenty times more expensive than overhead cables,[26] with topography, subsurface conditions, and opportunity to use existing underground conduit among the factors contributing to the differences in cost estimates. While underground lines are less prone to severe weather events, maintenance costs may be higher, especially if the line has to be unearthed or brought to the surface for repair, or if the area is prone to flooding. Thus, the perceived benefits of burying power lines sometimes comes at a cost which communities are not willing to accept.[27] Costs of undergrounding are usually passed along to electricity customers, and collected over an assumed average 25-year service life. Some utilities are also willing to share the cost with communities as they may derive some benefits from undergrounding.[28] However, plans to underground power lines are not always approved by regulatory bodies in some jurisdictions, as electric utilities and transmission builders are often required to consider "least cost" options.

Implementing Smart Grid Improvements

Much of the infrastructure which serves the U.S. power grid is aging. The average age of power plants is now over 30 years, with most of these facilities having a life expectancy of 40 years.[29] Electric transmission and distribution system components are similarly aging, with power transformers averaging over 40 years of age,[30] and 70% of transmission lines being 25 years old or older.[31] As components of the system are retired, they are replaced with newer components often linked to communications or automated systems.

The modernization of the grid to accommodate today's power flows, serve reliability needs, and meet future projected uses is leading to the incorporation of information processing capabilities for power system controls and operations monitoring. The "Smart Grid" is the name given to the evolving electric power network as new information technology systems and capacities are incorporated. A lot has been said about how the Smart Grid can potentially improve electric utility services, but many customers apparently have yet to be convinced that a Smart Grid will benefit them. Smart Grid improvements could help to avoid power outages, as the system would be able to detect problems, and potentially re-route power while alerting system operators to the location of the issue.[32]

Much of the Smart Grid would conceivably be deployed on electricity distribution systems. However, the privacy of customer information has been raised as an issue with smart meters (which relay information related to a customer's electricity usage back to the electric utility). Rate increases have accompanied Smart Grid-related projects in some states, as most of the costs related to smart meters have been directly passed on to electricity customers. Some state jurisdictions are requiring cost-benefit analyses before approving Smart Grid expenditures in rates.[33] Smart Grid programs will likely need to develop good performance metrics for reliability if these new capabilities are to provide the expected service improvements.[34]

Distributed Generation

Distributed generation is located close to the load it is meant to serve. According to DOE, distributed generation "... feeds into the distribution grid, rather than the bulk transmission grid."35 As such, distributed generation includes traditional back-up power sources (such as the large gas-powered generators used by institutions and companies), combined heat and power facilities (used for industrial, district, and community power generation), and renewable electricity power systems used by businesses and residences. Microgrids36 are another form of distributed generation. DOE estimated that as of 2007, there were more than 12 million distributed generation units installed across the United States, with a total capacity of 200 GigaWatts.[37]

Since distributed generation can directly serve the power consumer without external power lines, it is far less vulnerable to the kinds of weather-related service interruptions that can affect utility power systems. DOE stated that "weather is the primary reason for reliability problems, and includes problems caused by lightning strikes, high winds, snowfall, ice, and unexpectedly hot weather.

The goal of both planners and operators is to have as resilient a system as possible that can adjust to problems without causing major consequences, and that when outages do occur, they are short-lived and affect the fewest number of customers as possible."[38] Aside from the potential to provide back-up power to the customer in the event of an outage, DOE found there were other potential benefits to system reliability from distributed generation.[39]

- Several utilities offer financial incentives to owners of emergency power units to make them available to grid operators during times of system need.
- Several regions offer financial incentives or price signals to customers to reduce demand during times of system need (e.g., demand response programs), and some participants in these programs use distributed generation to maintain near-normal on-site operations while they reduce their demand for grid-connected power.

DOE further observed that distributed generation can add diversity to power supply options, thus leading to improvements in overall system adequacy. DOE also stated that distributed generation "has the potential to reduce the number of outages caused by overloaded utility equipment."[40]

Reliability-Centered Maintenance Regulations

A 2005 study[41] by Davies Consulting for the Edison Electric Institute (EEI) looked at state regulatory investigations of electric utility responses after major storm-related outages.[42] The study came to a number of conclusions concerning utility performance and infrastructure with regard to the time required to restore service after major storms:

- The systems are resilient.
- Adequate funds are being spent on maintenance and reliability.
- Restoration strategies, plans, and practices meet standard utility practices, but more effective resource acquisition (mutual aid) practices need to be employed, such as getting crews on the road earlier and allowing utilities to recover these costs even if they are not used.
- Communication processes with all stakeholders and customers need to be improved—before, during, and after an event.

- Restoration structures and emergency management structures and practices (e.g., Incident Command Structures) need to be aligned.

While the EEI-sponsored study found that adequate amounts were being spent in maintenance and reliability, the study suggested that reliability-centered maintenance[43] (RCM) practices may be an effective tool in preventing some storm-related outages, and could allow for a quicker recovery after a storm-related outage, and encouraged state utility regulators to consider approval of RCM if it was shown to be cost-effective.

> Regulatory efforts, with regard to maintenance, follow two general approaches: development of time-based maintenance intervals with compliance reporting, typically on an annual basis; and development of a higher level maintenance approach with assessments or self-certification. Time-based maintenance may be cost effective for some technologies, but it is likely not effective for all assets. As more utilities evaluate the RCM model, regulators will have to review the benefits and accept the model as part of a cost effective maintenance program. If RCM-type approaches are proven to be effective, but are not accepted by the commissions, utilities will be forced to continue to use the time based methodology and may face a "wall" where assets will need to be replaced in large quantities. As a result, utilities may begin a large-scale and unfocused replacement strategy that does not produce commensurate reliability.[44]

However, the study noted that some state jurisdictions have instituted quality-of-service benchmarks for electric utilities, with penalties imposed on utilities not achieving reliability performance targets. The study observed that penalties rather than rewards would be the focus, as "many state regulators feel that utilities should not be rewarded for service they should already be providing. Second, many regulators believe that there is no tangible benefit to most customers when a utility achieves performance beyond targets."[45] Additionally, those customers willing to pay a premium for reliability (the study reasoned) would likely install their own back-up or distributed generation.

Mutual Assistance Agreements

Depending on the size and extent of a storm causing an outage, a utility may find it necessary to call upon other utilities via its mutual assistance agreements[46] (MAAs) to help it restore services. MAAs can reduce the duration of weather-related outages by bringing in outside resources to aid the recovery effort.

> Mutual assistance is an essential part of the electric power industry's service restoration process and contingency planning. Electric utilities impacted by a major outage event are able to increase the size of their workforce by "borrowing" crews from other utilities. When called upon, a utility will send skilled line workers—both utility employees and contractors—along with specialized equipment to help with the restoration efforts of a fellow utility.
>
> ... Mutual Assistance includes primary and secondary response with back-up contingency plans. Contingency plans are developed by the smaller, more regionalized RMAGs (Regional Mutual Assistance Groups). These plans assure that resource transfers have a minimal effect on a regional area in case an unexpected event occurs. Contingency

plans can be done in various ways depending on the amount of resources being transferred either from a region or a larger geographic area such as across North America.47

MAAs are voluntary arrangements made with other electric utilities in states or regions unlikely to be affected by the same storm or weather-related event. As such, it may take time for MAAs to result in additional help arriving, as the utility workers must travel sometimes significant distances. Assuming that roadways are open, these convoys must travel at speeds which can be safely navigated by the bucket trucks and other specialized vehicles making the journey.

EEI provides advice on how to structure and arrange MAAs to its member utilities. While MAAs are voluntary and not regulated, they have come under state regulatory scrutiny especially in longer-duration outages. But the extent of damage can slow the response of other utilities to respond to calls for assistance, as a utility's first obligation is to restore service to its own customers in the event of an outage. Nevertheless, effective MAAs are invaluable to the recovery effort after a major storm-related outage.

CONCLUSION

In any discussion of storm-related power outages, two prominent themes emerge—preparation and recovery. If utilities are aware of an impending storm or weather-related event which may cause outages, they are expected to make preparations for restoration of services in as timely a manner as possible. Recovery from any such event will depend on the severity of the storm and the resulting damage. However, recovery can be hastened, and the amount of damage to electric power infrastructure can be minimized, if good maintenance, restoration, organization, and communications strategies are followed on an ongoing basis. Electric utilities should also be aware of local weather-related events in the past. If an area is prone to particular seasonal extremes of weather, then utilities would be expected to work with state and local regulatory authorities to look at ways to reduce vulnerabilities of the power and related infrastructure.

The cost of many aged parts of the electric power system have already been recovered in rates by electric utilities. The time for new investments may be approaching (if it is not already here) considering the projected future uses of the grid. Power delivery systems are most vulnerable to storms and extreme weather events. Improving the overall condition and efficiency of the power delivery system can only serve to improve the resiliency of the system, and help hasten recovery from weather-related outages. Ultimately, however, electric utilities are responsible for this infrastructure. They are in the business of selling electricity, and they cannot sell electricity if their power delivery systems are out of service.

Options for Congress

The inconsistency of data from outage reporting is an issue in quantifying the impacts of storm-related and other power outages. Congress could empower the National Institute of

Standards or DOE or some other federal agency to develop standards for the consistent reporting of power outage data.

While responsibility for the reliability of the bulk electric system is under FERC, no central responsibility exists for distribution systems. One possible option would be to bring these systems under the ERO for reliability purposes. However, this would require changes in law, with the Federal Power Act[48] being key to any changes.

Formalizing review or even coordinating MAAs under a federal agency like the Federal Emergency Management Agency or the Department of Homeland Security may be an option that Congress may want to consider. This would not necessarily mean federal approval of MAAs, but may help in the cooperative coordination of additional federal and state resources, especially in a wide, multi-state weather event. The utility industry, however, may be opposed to government involvement in MAAs.

The aged condition of the electrical grid in many parts of the United States will likely only detract from and compound recovery efforts. While there has been much discussion of transmission system inadequacies and inefficiencies, many distribution systems are in dire need of upgrades or repairs. The cost of upgrading the grid to meet future uses is expected to be high, as the estimates of the ASCE show, and the importance of a robust grid to U.S. economic performance goes without question. While the federal government recently made funding available for specific Smart Grid projects or transmission lines under the American Recovery and Reinvestment Act of 2009,[49] there has not been a comprehensive effort to study the needs, set goals, and provide targeted federal funding for the modernization of the U.S. grid as part of a long-term national energy strategy. Such an effort would also require decisions about the appropriate roles of government and the private sector.

End Notes

[1] A power outage results in a complete loss of power for the end-use customer.

[2] P.L. 109-58.

[3] 16 U.S.C. § 824o.

[4] Alison Silverstein, Transmission 101, National Association of Regulatory Utility Commissioners, April 20, 2011, p. 30, http://www.naruc.org/grants/Documents/ Silverstein%20NCEP%20T-101%20042011 1.pdf.

[5] Paul Hines, Jay Apt, and Sarosh Talukdar, Trends in the History of Large Blackouts in the United States, University of Vermont, 2008, http://www.uvm.edu/~phines/publications/ 2008/Hines_2008_blackouts.pdf.

[6] Kristina Hamachi LaCommare and Joseph H. Eto, Understanding the Cost of Power Interruptions to U.S. Electricity Consumers, Ernest Orlando Lawrence Berkeley National Laboratory, September 2004, http://certs.lbl.gov/pdf/ 55718.pdf.

[7] Per the definition of a "sustained interruption" used by the Institute of Electrical and Electronics Engineers (IEEE). Guide for Electric Power Distribution Reliability Indices, No. 1366.

[8] "... widespread power losses resulting from major natural events (primarily storms but also hurricanes and earthquakes) are sometimes not included in the same data categories as more routine power losses. As a result, power losses from natural events are not always included in data used for cost estimates." LaCommare and Eto, op. cit., p. 5.

[9] U.S. Department of Energy, Energy Information Administration, Form OE-417. See http://www.eia.gov/cneaf/ electricity/page/disturb_events.html.

[10] Evan Mills, Extreme Grid Disruptions and Extreme Weather, Lawrence Berkeley National Laboratory, U.S. DisasterReanalysis Workshop, May 3, 2012, http://evanmills.lbl.gov/ presentations/Mills-Grid-Disruptions-NCDC3May2012.pdf.

[11] E-mail from Evan Mills, Lawrence Berkeley National Laboratory, August 30, 2012.

[12] Council of European Energy Regulators, 4th Benchmarking Report on Quality of Electricity Supply 2008, C08-EQS24-04, December 10, 2008, http://www.energy-regulators.eu/ portal/page/portal/EER_HOME/EER_PUBLICATIONS/ CEER_PAPERS/Electricity/2008/C08-EQS-24-04_4th%20Benchmarking%20 Report%20 EQS_10-Dec-2008_re.pdf.

[13] Ibid.

[14] Ibid, p. 9.

[15] Ibid, page 7.

[16] "The cumulative need, based on anticipated investment levels and the estimated investment gap, will be $673 billion by 2020, an average of about $75 billion per year. Based on investment over the past decade, closing the gap is within reach: the average annual need projected from 2012 through 2020 falls within the range of annual investment totals in the last decade, and there is not a single year through 2020 that is projected to be outside that range." American Society of Civil Engineers, Failure to Act: The Economic Impact of Current Investment Trends in Electricity Infrastructure, 2011, http://www.asce.org/ uploadedFiles/Infrastructure/Failure_to_Act/energy_report_FINAL2.pdf.

[17] "Based on current investment trends, the national electricity infrastructure gap is estimated to be $107B by 2020, or just over $11B per year. By 2020, shortfalls in grid investments are expected to account for almost 90% of the investment gap with nearly $95B in additional dollars needed to modernize the grid." Ibid.

[18] Consortium for Electric Infrastructure for a Digital Society, The Cost of Power Disturbances to Industrial & Digital Economy Companies, June 29, 2001, http://www.onpower. com/pdf/EPRICostOfPowerProblems.pdf.

[19] "The market value of goods and services produced by labor and property in the United States, regardless of nationality." See http://www.bea.gov/glossary/glossary_g.htm.

[20] Consortium for Electric Infrastructure for a Digital Society, op. cit., p. 16.

[21] LaCommare and Eto, op. cit., Executive Summary.

[22] A strip of land occupied by active electricity transmission or distribution facilities.

[23] North American Electric Reliability Corporation, Transmission Vegetation Management NERC Standard FAC-003-2 Technical Reference, September 2009, http://www.nerc. com/docs/standards/sar/FAC-003-2_White_Paper_2009Sept9.pdf.

[24] Clearances at sea level. Ibid, page 47.

[25] Lee R. Hansen, Utility Tree Trimming in Other States, State of Connecticut, December 19, 2011, http://www.cga.ct.gov/2011/rpt/2011-R-0459.htm.

[26] Jeff Griffin, "Underground Electric Transmission Installations Gaining Traction," Underground Construction, vol. 65, no. 6, June 2010, http://www.underground construction magazine.com/underground-electric-transmissioninstallations-gaining-traction.

[27] Julie Patel, "Are You Willing to Pay Thousands to Keep the Power on After a Storm?," Sun Sentinel, June 25, 2011, http://articles.sun-sentinel.com/2011-06-25/business/fl-fpl-underground-lines-20110624_1_power-lines-utility-linestamara-tennant.

[28] Ken Silverstein, "Upgrading the Grid by Going Underground," Forbes, August 2, 2012, http://www.forbes.com/sites/ kensilverstein/2012/08/02/upgrading-the-grid-by-going-under ground/2/.

[29] Massachusetts Institute of Technology, Retrofitting of Coal-Fired Power Plants for CO2 Emissions Reductions, March 23, 2009, http://web.mit.edu/mitei/docs/reports/meeting-report.pdf.

[30] Thomas A. Prevost and David J. Woodcock, Transformer Fleet Health and Risk Assessment, Weidman Electrical Technology, IEEE PES Transformers Committee Tutorial, March 13, 2007, http://grouper.ieee.org/groups/transformers/info/S07/S07-TR_LifeExtension.pdf.

[31] K. Anderson, D. Furey, and K. Omar, Frayed Wires: U.S. Transmission System Shows Its Age, Fitch Ratings, October 25, 2006.

[32] Ken Silverstein, op. cit.

[33] Katherine Tweed, Illinois Rejects Ameren's Smart Grid Plan, Greentech Media, May 30, 2012, http://www.greentechmedia.com/articles/read/illinois-rejects-amerens-smart-grid-plans1/.

[34] Galvin Electricity Initiative, Electricity Reliability: Problems, Progress and Policy Solutions, February 2011, http://www.galvinpower.org/sites/default/files/Electricity_ Reliability_ 031611.pdf.

[35] U.S. Department of Energy, The Potential Benefits of Distributed Generation and Rate-Related Issues That May Impede Their Expansion, February 2007, http://www.ferc. gov/legal/fed-sta/exp-study.pdf.

[36] A microgrid is any small or local electric power system that is independent of the bulk electric power network. For example, it can be a combined heat and power system based on a natural gas combustion turbine (which cogenerates electricity, and hot water or steam from water used to cool the natural gas turbine), or diesel generators, renewable energy, or fuel cells. A microgrid can be used to serve the electricity needs of data centers, colleges, hospitals, factories, military bases, or entire communities (i.e., "village power").

[37] "Over 99% of these units are small emergency reciprocating engine generators or photovoltaic systems, installed with inverters that do not feed electricity directly into the distribution grid. However, ... this large number of smaller machines represents a relatively small fraction of the total installed capacity." U.S. Department of Energy, The Potential Benefits of Distributed Generation and Rate-Related Issues That May Impede Their Expansion, February 2007, http://www.ferc.gov/legal/fed-sta/exp-study.pdf, p. 33.

[38] U.S. Department of Energy, The Potential Benefits of Distributed Generation and Rate-Related Issues That May Impede Their Expansion, February 2007, http://www.ferc.gov /legal/fed-sta/exp-study.pdf, p. 42.

[39] Ibid.

[40] Ibid. p. 50.

[41] Davies Consulting, Inc., State of Reliability Distribution Regulation in the United States, Edison Electric Institute, September 2005, http://legalectric.org/f/2010/04/ stateof distributionreliability-2005.pdf.

[42] The events included the North Carolina and South Carolina ice storm of December 2003, Hurricanes Isabel and Juan (September 2003), the Utah snow storm of December 2004, four hurricanes in Florida in summer 2004, and the Maritime ice storm of November 2004.

[43] "Reliability centered maintenance ... is a scientific process designed to develop a maintenance program, focuses on equipment condition and is matched to the wear-out pattern of that equipment based on operation, not time. Those industries using reliability centered maintenance such as the U.S. airline industry, the U.S. Navy (including nuclear submarines), nuclear power plants, and fossil fuel plants, indicate that adoption of RCM has increased reliability while simultaneously reducing costs." Davies Consulting, Inc., op. cit., p. 21.

[44] Ibid.

[45] Ibid.

[46] Most utilities have found that they cannot staff up internally to provide the manpower they need to respond to major storms. Therefore, they rely on mutual assistance agreements where they agree to share line crews and equipment to help each other respond to major storms. See http://www.eei.org/ourissues/electricitydistribution/ Documents/ StormRestore Report.pdf.

[47] See http://www.eei.org/ourissues/ElectricityTransmission/ Reliability/Pages/ElectricSector MutualAssistance. aspx.

[48] 16 U.S.C. § 824o.

[49] The American Recovery and Reinvestment Act of 2009 (P.L. 111-5) made available $3.4 billion for funding Smart Grid projects, and approximately $6.5 billion in borrowing authority for transmission systems was authorized for the Western Area Power Administration and the Bonneville Power Administration. Additionally, $6 billion in loan guarantees was made available for renewable energy and electric power transmission.

In: The Transformation of Electrical Power: Key Issues
Editor: Dawson B. Munoz
ISBN: 978-1-62618-619-4

Chapter 6

EPA'S UTILITY MACT: WILL THE LIGHTS GO OUT?*

James E. McCarthy

SUMMARY

On December 21, 2011, EPA Administrator Lisa Jackson announced final standards aimed at reducing mercury and other air toxics emissions from electric generating units (EGUs) by about 90%. The rule, commonly referred to as the "Utility MACT" or the "Mercury and Air Toxics Standards" (MATS), has been more than a decade in the making (Congress authorized the standards in the 1990 Clean Air Act Amendments), and it is among the most expensive rules that EPA has ever promulgated. EPA estimates the annualized cost at $9.6 billion in 2015. Industry estimates have been higher.

The benefits are also large, according to EPA, ranging from $37 billion to $90 billion annually. The benefits mostly reflect the monetized value of avoiding up to 11,000 premature deaths annually.

The rule's costs will fall primarily on older coal-fired units that do not have state-of-the art pollution controls. EPA says that this is a minority of coal-fired plants and an even smaller share of all electric generation: the agency estimates that 56% of coal-fired units have already installed equipment that can be used to meet the standards. In addition, about 55% of the nation's electricity supply comes from natural gas, nuclear, and renewable sources that are not subject to the rule's requirements.

This report describes the rule and its potential impact. The report discusses previous EPA efforts to regulate utility mercury emissions, the court decision overturning those regulations, the specifics of the new rule, its estimated costs and benefits, the impact of the rule on electric reliability, and legislation related to it that has been or may be considered in the 112th Congress. Industry and environmental groups have been keenly interested in both the substance of the rule and the schedule for its implementation, and the House has already

* This is an edited, reformatted and augmented version of the Congressional Research Service Publication, CRS Report for Congress R42144, dated January 9, 2012.

passed legislation (H.R. 2401) that would change both. A particular issue has been whether the standards will lead to retirement of a significant number of electric generating units, with negative effects on the reliability of the power supply. EPA and many other analysts maintain that this will not be the case.

To address this question, this report reviews industry data on planning reserve margins and potential retirement of units that do not currently meet the standards. Based on these data, it appears that, although the rule may lead to the retirement or derating of some facilities, almost all of the capacity reductions will occur in areas that have substantial reserve margins. Two areas that may have difficulty meeting reserve margins, Texas and New England, will experience few plant retirements and deratings, according to industry data. Furthermore, to address the reliability concerns expressed by industry, the final rule includes provisions aimed at providing additional time for compliance if it is needed to install pollution controls or add new capacity to ensure reliability in specific areas.

As a result, it is unlikely that electric reliability will be harmed by the rule. Another potential concern, given the rule's cost, is what impact it may have on the price of electricity. EPA estimates that the average price of electricity nationally will increase by 3.1% by 2015, as a result of the rule. Electricity prices have declined more than 20% in real terms since 1980. The impact of price changes would be relatively small compared to this downward trend, and well within the normal range of historical price fluctuations.

INTRODUCTION

On December 21, 2011, EPA Administrator Lisa Jackson announced final standards for mercury and other air toxics emissions from electric generating units (EGUs).[1] The rule, commonly referred to as the "Utility MACT" or the "Mercury and Air Toxics Standards" (MATS), has been a long time in the making: Congress authorized the standards in the 1990 Clean Air Act Amendments, and EPA made a preliminary determination and began developing actual standards in 2000.

The rule is among the most expensive rules that EPA has ever promulgated: EPA estimates the annualized cost at $9.6 billion in 2015.[2] Industry estimates have been higher,[3] although most analyses were conducted before EPA proposed or promulgated the final rule. Industry and environmental groups have been keenly interested in both the substance of the rule and the schedule for its implementation, and the House has already passed legislation (H.R. 2401) that would change both.

This report describes the rule and its potential impact. The report begins with a background section that describes the statutory authority and history of the rule's development before discussing the specifics of the rule, its estimated costs and benefits, and legislation related to it that has been considered in Congress.

BACKGROUND

In 1990, when Congress enacted amendments to the Clean Air Act, one of the major changes to the act was a new set of requirements that EPA was to impose on sources of hazardous air pollutants. Unhappy with the slow pace at which EPA was regulating emissions of these pollutants, Congress removed most of the agency's discretion as it restructured the program: it listed 189 hazardous air pollutants (HAPs) in the statute[4] and required EPA to identify sources that emitted more than 10 tons of any individual HAP or 25 tons of any combination.[5] EPA was directed to impose the "Maximum Achievable Control Technology" on these sources, a term defined with great precision in the statute. Standards for all categories and subcategories of sources were to be promulgated no later than 10 years after the date of enactment (i.e., by November 15, 2000), and sources were to comply with the standards three years after their promulgation, with one additional year available in limited cases.

Electric utilities lobbied hard for an exception to these requirements. What they got was a special subsection of the act (Section 112(n)(1)) that required EPA to undertake a study and report to Congress on the hazards to public health reasonably anticipated to occur as a result of emissions by EGUs and alternative control strategies for the emissions. The study was to be completed within three years of the date of enactment, and the EPA Administrator was directed to regulate EGUs if she found such regulation to be "appropriate and necessary after considering the results of the study."

The study was completed in February 1998,[6] and the Administrator published a finding that regulation was appropriate and necessary on December 20, 2000.[7]

Although EGUs emit numerous HAPs, the principal HAP of concern has been mercury. Mercury is a potent neurotoxin that can harm health at very low concentrations—the principal effects being delayed development, neurological defects, and lower IQ in fetuses and children.[8] The principal route of exposure to mercury is through consumption of fish. Mercury enters water bodies, often through air emissions, and is taken up through the food chain, ultimately affecting humans as a result of fish consumption. All 50 states have issued fish consumption advisories due to mercury pollution, covering 16.8 million acres of lakes, 1.25 million river miles, and the coastal waters of 20 entire states.[9] Coal-fired electric generating units account for about half of U.S. mercury emissions.

Based largely on the concerns about mercury, the Administrator determined that regulation was appropriate and necessary in December 2000, and the agency proceeded to develop regulations for the EGU category.

In 2005, however, upon completion of the process, EPA promulgated regulations establishing a cap-and-trade system to limit emissions of mercury from coal-fired power plants rather than the plant-specific Maximum Achievable Control Technology (MACT) standards required by Section 112 of the act. All previous sources of mercury subject to emission standards had been required to meet plant-specific MACT standards.[10]

Section 112 sets out very detailed requirements for MACT standards, including a list of the pollutants that need to be controlled (not just mercury, but any of 186 other HAPs) and the level of control that the standards must achieve.

The 2005 cap-and-trade rules addressed only mercury, and would have allowed many power plants to avoid control provided they obtained allowances from others who achieved lower pollution levels than required, or reduced emissions sooner than required. The ability of plants to avoid emission control by purchasing allowances could lead to the continuation of "hot spots," areas where mercury concentrations in waterbodies are greater than elsewhere.

By contrast, MACT standards are required by Section 112 of the act to be applicable at each existing plant and to be no less stringent than the average emission limitation achieved by the best performing 12% of existing sources in the industry subcategory.[11] These statutory requirements are referred to as the "MACT floor," because the agency is not allowed to set less stringent standards, nor may it take economic factors into account in determining what the floor will be.

Whether the agency could substitute cap-and-trade rules for the MACT requirements was challenged by the State of New Jersey and others, and, in a 3-0 decision, the D.C. Circuit Court of Appeals vacated the cap-and-trade rules in 2008.[12]

Rather than appeal the court's ruling to the Supreme Court or attempt to delist the category, EPA proposed the Utility MACT standards on March 16, 2011, and finalized them on December 21. In between those dates, the agency held three public hearings and provided a 90-day public comment period that resulted in 960,000 comments on the rule, of which the agency said approximately 20,000 were unique comments.

As of early January 2012, the final rule had not appeared in the Federal Register, but a pre-publication copy is available at http://www.epa.gov/airquality/ powerplanttoxics/pdfs/ 20111216MATSfinal.pdf.

THE FINAL RULE

The final rule is similar to the March 2011 proposal.[13] The rule will result in about a 90% reduction from uncontrolled power plant emissions of mercury, nine other toxic metals, and three acid gases, all of which were listed by Congress as hazardous air pollutants in the 1990 Clean Air Act Amendments.[14]

According to EPA, power plants are the largest emitters of many of these pollutants, accounting for about 50% of the nation's mercury emissions, 62% of arsenic emissions, and 82% of hydrochloric acid emissions, for example.[15] The Utility MACT will also reduce emissions of sulfur dioxide and fine particulates ($PM_{2.5}$), which, although not categorized as hazardous air pollutants, are estimated to cause thousands of premature deaths annually.

The rule affects about 1,100 coal-fired and 300 oil-fired units.[16] Despite the fairly large number of oil-fired units, they account for only 1% of U.S. electricity generation, and they are mostly unaffected by the rule. EPA estimates that the total annualized costs for compliance by oil-fired units will be $56 million (less than 0.1% of the rule's cost), and that there will be no change in oil-fired capacity as a result of the rule.[17] Thus, the remainder of this discussion focuses on coal-fired units, which will bear the brunt of the compliance cost.

In proposing the standards, EPA noted that while the requirements are stringent for those facilities lacking controls, 56% of existing coal-fired power plants already have equipment in place needed to achieve compliance.

Thus, when implemented in 2015 or 2016, the standards are expected to level the playing field, bringing older, poorly controlled plants up to the standards being achieved by a majority of the existing units. In this respect, the standards reflect the statute's requirement that existing sources of HAPs should meet standards based on the current emissions of the best performing similar sources.

EPA's Analysis of Costs, Benefits, and Control Technology

Costs

EPA projects the annualized cost of compliance with the final MATS rule at $9.6 billion in 2015, gradually declining to $7.4 billion annually in 2030. Total electric utility revenues from sales to ultimate customers equaled $369 billion in 2010, according to the Energy Information Administration (EIA).[18] Thus, the annualized cost would be less than 3% of current industry revenues. For individual companies and plants, however, and for regions of the country, the percentage may be higher or lower depending on the type of units operated by the company (coal, gas, nuclear, etc.) and, for coal-fired units, whether they have already installed control equipment. The average consumer will see an increase of 3.1% ($3-$4 per month) in the cost of electricity in 2015 due to the rule, according to EPA, falling to less than 1% by 2030.[19] As shown in **Figure 1**, electricity prices have declined by about 20% since the early 1980s, after adjusting for inflation.

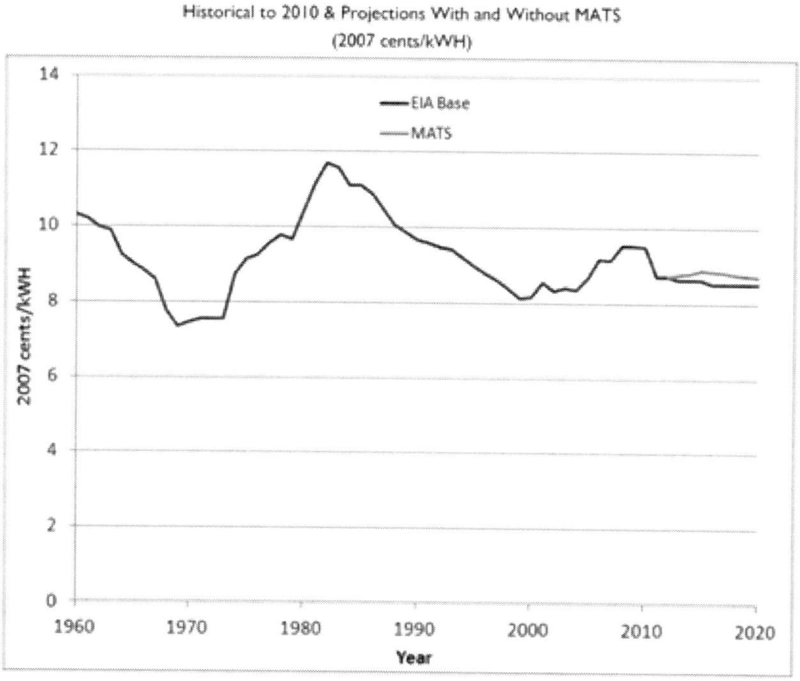

Source: U.S. EPA. EIA base refers to historical data and projections of the U.S. Department of Energy's Energy Information Administration.

Figure 1. Average U.S. Total Electricity Prices.

Reflecting the range of cost increases experienced by utilities, EPA estimates that retail electricity price increases will also vary, from a low of 1.3% in California, to a high of 6.3% in the area served by the Southwest Power Pool (SPP)—Oklahoma, Kansas, and parts of five other states. The SPP region currently has the second lowest electricity prices in the country, and will continue to have the second lowest prices even after increases due to the MACT rule, according to EPA.[20]

Control Technology

The cost of compliance with the Utility MACT/MATS rule results largely from the installation of scrubbers and fabric filters on coal-fired generation units.[21] As a result of the rule, 20 gigawatts (GW) of coal-fired units, about 7% of total coal-fired capacity, are expected to install scrubbers. EPA estimates that by the time the rule requires compliance, 203 GW will already have installed scrubbers as a result of other regulations, so the new installations, while expensive, are more incremental rather than a major departure from current practice. About one-third of the already installed scrubbers (63 GW) will need to be upgraded to improve their ability to capture pollutants.

One-third of the coal-fired EGU capacity (102 GW) are expected to add fabric filters because of the rule, while 90 GW would have them whether or not there were a rule. In most cases, the fabric filters will be coupled with activated carbon injection or dry sorbent injection. Mercury and other HAPs become attached to the carbon or sorbent after it is injected into the flue gas, and the fabric filter collects the particles, removing them from the plant's emissions. EPA estimates that 99 GW of coal-fired capacity (about one-third of the U.S. total) will add activated carbon injection by 2015 because of the rule. The rule also results in 44 GW of dry sorbent installations.[22]

This is not complicated or new technology. Other types of facilities (notably solid waste incinerators) have used this technology for the past 15 years to reduce their mercury and other HAP emissions by 95% or more. As a result of state-level pollution control regulations, a growing percentage of coal-fired power plants do the same.

Benefits

The benefits of the rule are estimated by EPA at $37 billion to $90 billion annually—4 to 9 times as great as the costs—due primarily to the avoidance of up to 11,000 premature deaths each year. Other benefits, only some of which were given dollar values, include the annual avoidance of 4,700 nonfatal heart attacks, 130,000 asthma attacks, and developmental effects on children, including effects on IQ, learning, and memory.[23]

Economic Impacts

Of the dozens of recently proposed EPA rules, the Utility MACT is probably the most costly.[24] It is likely to affect older coal-fired plants that have not yet installed current pollution control technology. The agency concluded that some of these plants, representing less than 5 GW of coal-fired capacity, would be retired by 2015, rather than invest in control technologies. In all, it said, coal-fired generation would decline 1.3% compared to estimated generation in the absence of the rule.[25]

Retirement of these older plants could lead to job losses at some specific locations, but overall, EPA's analysis concludes that the rule is likely to lead to an increase in employment. The agency finds that 46,000 job-years will be required for construction/installation of pollution control equipment and 8,000 long-term utility jobs will be created to operate and maintain the controls.[26] Others challenge these conclusions: the Edison Electric Institute, the National Association of Manufacturers, and other industry groups maintain that increased costs of electricity will cause job losses in industries that rely heavily on electricity and are located in areas of the country that will be most affected by the rule.

Although generally requiring controls on each individual source, the regulations do allow averaging of emissions from multiple units at a single location, which may allow some older units that are operated infrequently to remain in service. The absence of broader allowance trading provisions in the authorizing statute and the stringency of the emission requirements mean that most units will not be able to escape regulation.

INDUSTRY ANALYSES OF THE UTILITY MACT RULE

Electric power generators are split over EPA's rule. Companies that rely on nuclear power or natural gas for most of their power, and have fewer coal-fired plants, and companies that have already invested in controls due to state requirements or other federal regulations, generally support the rule. A number of these utilities, accounting for about 20% of U.S. electric generating capacity, have formed the Clean Energy Group. At an April hearing of the House Energy and Commerce Committee, Michael Bradley, the group's Director, stated:

> While not perfect, the proposal is reasonable and consistent with the requirements of the Clean Air Act.... While complying with these obligations will take planning and significant resources by the electric sector, many companies are well on their way toward compliance and, based on the proposed rule, we anticipate that the electric sector can comply with the Act's requirements. There is no reason to delay the implementation of the Utility Toxics Rule.[27]

Opponents of the rule have the support of the industry's main trade association, the Edison Electric Institute (EEI), as well as groups representing the coal industry. In a report written before EPA even proposed the Utility MACT, EEI concluded that, "All coal units [would be] required to install a scrubber (wet or dry), activated carbon injection (ACI) and a baghouse/fabric filter" for compliance with the MACT.[28] This goes well beyond what EPA proposed and what it promulgated in the final rule. Compared to EPA's projections, EEI concluded that five times as much scrubber capacity, nearly three times as much ACI, and about one and one-half times as much baghouse capacity would need to be added, making the rule substantially more costly and far more difficult to comply with in the limited time provided by the statute.

A report by the North American Electric Reliability Corporation (NERC),[29] which was also written before EPA proposed the Utility MACT, also assumed that vastly more pollution control equipment would need to be added to coal-fired plants than EPA believes will be necessary. The NERC analysis assumed wet scrubbers would be added to all coal-fired plants that don't already have them, that selective catalytic reduction (SCR) will be added to all bituminous coal-powered facilities, and that activated carbon injection and baghouses/fabric filters would be added at all facilities burning other types of coal.[30] These assumptions are similar to EEI's except that by assuming wet scrubbers (instead of EPA's general assumption that dry scrubbers will suffice) and by assuming SCR at bituminous facilities, the cost impacts would most likely be even greater than the costs in EEI's assessment.[31]

A November 2011 Reliability Assessment by NERC made minor adjustments to these assumptions, but still concluded that scrubbers would be required on all operating coal-fired facilities and that a much higher percentage of facilities would require baghouses and ACI than EPA projected. Despite these more costly assumptions, NERC found that generation capacity would decline only about 1% nationwide as a result of the Utility MACT.[32] The 2011 report concludes that 6.5 to 9.9 GW of coal-fired capacity would be retired and 2.1 GW would be derated[33] as a result of the MACT rule.

WILL THE LIGHTS GO OUT?

NERC also looked at reserve margins in 18 regions covering the 48 mainland U.S. states, and found that only two of the regions – ERCOT (the Texas electrical grid, which is not interconnected with other regional transmission organizations) and New England – would experience planning reserve margins below the NERC Reference Level of 15% in 2015.[34] See **Figure 2**.

In both cases, the low reserve margins appear to have little to do with the Utility MACT or other EPA rules. In 2015, the ERCOT region will have 79,415 megawatts (MW) of anticipated generation capacity, giving it an anticipated reserve margin of 10.4%. Deratings and retirements due to EPA rules will have reduced that capacity by only 267 MW (0.34%), according to NERC.[35] The NERC data suggest that ERCOT may experience reliability problems, but the emission standards will play a minor role.

In New England, NERC reports conflicting information. It begins its analysis by stating:

> For this *2011 Long-Term Reliability Assessment*, ISO New England Inc. (ISO-NE) forecasts no major reliability issues with respect to fuel supply, availability of both supply or demand-side resources, or the capability of the regional transmission system to serve the projected seasonal peak demands and energy requirements of the six-state New England subregion.[36]

Following this statement, however, the report shows reserve margins consistently below 15%. As with Texas, the low reserve margins would exist even in the absence of the Utility MACT. NERC finds that the new EPA regulations will cause 237 MW of retirements or deratings, 0.7% of the region's 32,649 MW of anticipated capacity.

According to NERC, the EPA rule will have its biggest impact in three regions known as PJM, MISO, and SERC-SE.[37] These areas cover 12 states in the Midwest and Mid-Atlantic, plus Alabama, Georgia, the District of Columbia, and small parts of four other states.[38] In SERC-SE, the area that covers Georgia and Alabama, the retirements and deratings remove 2.7% of anticipated capacity. But none of these areas falls below a 17% reserve margin.

Time Needed to Comply

Following publication of the final standards in the *Federal Register*, existing power plants will have three years, with a possible one-year extension, to meet the standards. (The three-to-four-year timeframe is mandated by the statute.) Many in industry argue that three or four years is not enough time to complete installation of the required pollution control equipment, and as a result that the reliability of the nation's electric power supply could be harmed by the rule while construction is ongoing, even if ultimately the industry will have sufficient generating capacity. NERC did not say this directly, in part because its analysis combines the effects of four rules, making it difficult to disaggregate the Utility MACT's effect. What it did say was:

> To comply [with the MACT Rule], owners of the remaining capacity need to retrofit from 277 to 753 units with added environmental controls. The "hard stop" 2015 compliance deadline proposed by the MACT Rule makes retrofit timing a significant issue and potentially problematic.[39]

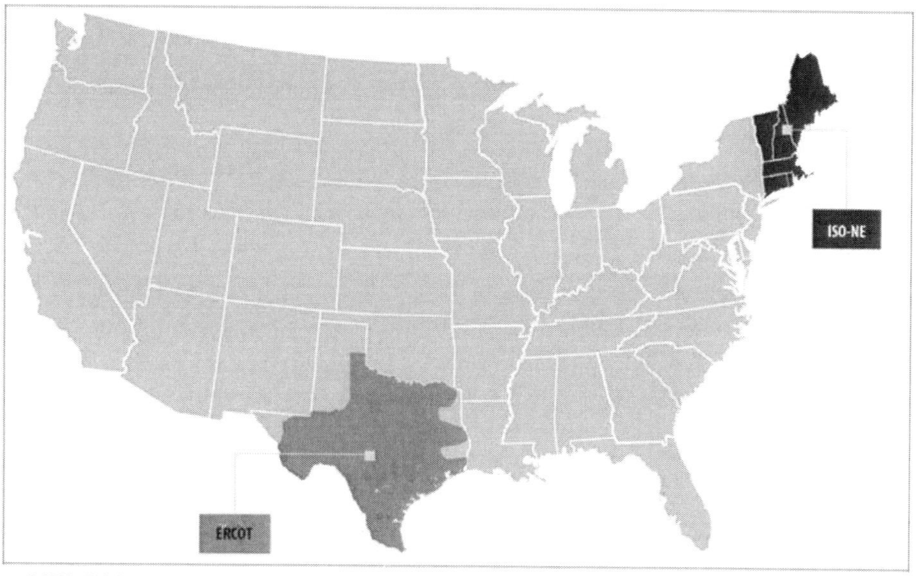

Source: NERC 2011 Long-Term Reliability Assessment.

Notes: 1. Both ERCOT and ISO-NE are projected to have reserve margins below 15% whether or not EPA implements the MATS rule. 2. ISO-NE (New England) forecasts no major reliability issues, according to NERC, despite having anticipated reserve margins below 15% under some assumptions. 3. All other areas have anticipated reserve margins above 15% with or without MATS.

Figure 2. Planning Areas with Anticipated Reserve Margins Below 15% in 2015.

Whether or not there is sufficient time to implement the rule without threatening electric system reliability will depend, to some extent, on the number of units that require retrofits. Both the EEI and NERC analyses discussed above assumed requirements that appear to be substantially more stringent than what EPA has promulgated. If EPA is correct in its analysis of how EGUs can comply with the rule, the number of retrofits appears to be within the range of what the industry has accomplished in the past as a result of earlier regulations.

Although EPA believes that most units will be able to comply with the Utility MACT within the three-year statutory deadline, in response to industry comments on the proposed rule, EPA and the White House added provisions to the final rule that will make additional time available for compliance where necessary. The final rule was accompanied by a Presidential memorandum directing the agency to make full use of the Clean Air Act's authority to grant additional time to those facilities that need it.[40] In general, this means that the agency will approve state permitting actions that provide an additional year (four years instead of three) for compliance if it is "necessary for the installation of controls." In the Preamble to the final rule, EPA discusses at some length its interpretation of the phrase "necessary for the installation of controls," making it clear that it intends to broadly interpret the phrase in order to make additional time available where needed.[41]

EPA also issued an enforcement policy that describes how units may obtain a fifth year for compliance under the agency's authority in Section 113 of the Clean Air Act.[42] The policy states:

> The EPA generally does not speak publicly to the intended scope of its enforcement efforts, particularly years in advance of the date when a violation may occur. The Agency is doing so now with respect to the MATS to provide confidence with respect to electric reliability....
>
> Some sources may take all steps necessary to comply with the MATS, but may nevertheless be needed to operate in noncompliance with the MATS to address concerns with electric reliability. In the event that such sources are interested in receiving a schedule to come into compliance while operating, the EPA intends, where necessary to avoid a serious risk to electric reliability, and provided the criteria set forth herein are met [these criteria essentially require timely notice to EPA and the appropriate planning authorities], to issue an expeditious case-specific AO [Administrative Order] to bring a source into compliance within one year.[43]

LEGISLATION

As noted earlier, the proposed Utility MACT drew an extraordinarily large number of public comments. While many of these were duplicative, the agency said that it received about 20,000 unique comments.

Congress has also taken a keen interest in the proposal. H.R. 2401, the Transparency in Regulatory Analysis of Impacts on the Nation (TRAIN) Act of 2011, which the House passed September 23, has been one congressional response. The bill would establish a panel of representatives from 11 federal agencies to report to Congress by August 2012 on the cumulative economic impact of the Utility MACT and a number of other listed EPA rules, guidelines, and actions concerning clean air and waste management. It would render the Utility MACT and another rule that EPA has promulgated to address power plant emissions (the Cross-State Air Pollution Rule) "of no force and effect"; it would delay promulgation of a replacement for the Utility MACT until at least one year after submission of the cumulative impacts report and delay compliance for at least five years after that date; and it would require that the replacement rule impose the least burdensome regulatory alternative from among the alternatives authorized under the Clean Air Act, among other provisions.

The Senate has not acted on H.R. 2401. A Senate bill, S. 609, similar to an earlier version of H.R. 2401, has also seen no action. For the Senate to act on either bill would generally require a favorable report from the committee of jurisdiction (the Senate Environment and Public Works Committee) and the votes of 60 Senators in order to get the bill considered on the floor.

There is another route, however, that removes the main obstacles to Senate action on the regulations: the Congressional Review Act (CRA). Under the CRA, after a rule is final and is transmitted to Congress, the Congress can consider a resolution of disapproval under special procedures. If a CRA resolution disapproving a rule is enacted, the rule cannot take effect, and the agency may not reissue either that rule or any substantially similar one, except under authority of a subsequently enacted law. It is widely expected that CRA resolutions disapproving the Utility MACT will be introduced during the second session of the 112th Congress.

What is unique about CRA resolutions is that the act provides special procedures to make it easier to obtain a vote in the Senate. CRA resolutions come to the Senate floor if 30 Senators sign a discharge petition within a specified period of time. Once a disapproval resolution is on the Senate Calendar, a motion to proceed to consider it is in order; the motion is not debatable and cannot be filibustered through extended debate. Debate is limited to 10 hours and amendments are prohibited.

Although it may be easier to obtain congressional approval of a CRA resolution, the path to enactment of such a resolution is still a steep one. The Obama Administration has made a significant commitment to promulgation of the Utility MACT and considers it one of the Administration's major achievements. As a result, legislation restricting EPA's authority to act, if passed by Congress, would likely encounter a presidential veto. There are no special procedures for Senate consideration of a CRA veto override. Overriding a veto (whether for a resolution of disapproval or other legislation) requires a two-thirds majority in both the House and Senate, and is seen by many as unlikely.

Unless the Presidency changes hands in the next election, that would leave the courts as the most likely venue for a successful challenge to the rule. A court challenge can take years. The challenge to EPA's 2005 utility mercury rule, for example, took nearly three years before the D.C. Circuit Court of Appeals overturned it. In the meantime, unless stayed by the court, the rule will be in effect, and power companies, facing a tight deadline, will need to begin the planning, design, and construction necessary to comply with the standards.

End Notes

[1] The rule has not yet been published in the *Federal Register*. A pre-publication copy of the rule and explanatory materials are available on EPA's website at http://www.epa.gov/ airquality/powerplanttoxics/actions.html.

[2] U.S. EPA, Office of Air Quality Planning and Standards, *Regulatory Impact Analysis for the Final Mercury and Air Toxics Standards*, December, 2011, p. 3-31, at http://www.epa.gov/ ttn/ecas/regdata/RIAs/matsriafinal.pdf. Hereinafter, "Regulatory Impact Analysis."

[3] For a discussion of the impacts of the Utility MACT and other rules affecting the electric power industry, see CRS Report R41914, *EPA's Regulation of Coal-Fired Power: Is a "Train Wreck" Coming?* One cost estimate issued after the rule's release can be found at National Association of Manufacturers, "Manufacturers: Utility

MACT Is Extremely Costly Regulation," Press Release, December 21, 2011, at http://www.nam.org/Communications/ Articles/2011/12/Manufacturers-Utility-MACT-Is-Extremely-Costly-Regulation.aspx.

[4] The list is found in Section 112(b) of the act. Congress and EPA have each removed one HAP from the list since 1990, leaving 187 HAPs on the current list.

[5] Overall, EPA identified 174 industrial categories to be regulated under the MACT provisions. Standards have been promulgated for almost all these categories except EGUs.

[6] The study is available at http://www.epa.gov/ttn/oarpg/t3/reports/eurtc1.pdf.

[7] U.S. EPA, "Regulatory Finding on the Emissions of Hazardous Air Pollutants from Electric Utility Steam Generating Units," 65 *Federal Register* 79825, December 20, 2000. In Section III of the Preamble to the final (December 2011) MATS standards, the Administrator affirmed the "appropriate and necessary" finding. A pre-publication copy is available at http://www.epa.gov/airquality/powerplanttoxics/pdfs/ 20111216 MATSfinal.pdf.

[8] For a detailed discussion of mercury's health effects, see CRS Report RL32420, *Mercury in the Environment: Sources and Health Risks.*

[9] See U.S. EPA, "National Listing of Fish Advisories: Technical Fact Sheet," September 2009, at http://water.epa.gov/scitech/swguidance/fishshellfish/fishadvisories/tech2008.cfm.

[10] Besides EGUs, the two biggest U.S. sources of mercury in 1990 were municipal waste incinerators and medical waste incinerators. EPA regulations required both of these categories to meet MACT standards in the 1990s, and they reduced their mercury emissions by 96% and 98% respectively by 2005. EGU emissions, not subject to MACT standards, were only reduced 10% during that period.

[11] For new sources, the standards are to be based on the emission control achieved by the best controlled similar source.

[12] New Jersey v. EPA, 517 F.3d 574 (D.C. Cir. 2008). The court found that, under Section 112, unless EPA "delisted" the category of HAP sources, it had to require that each plant in the category meet MACT standards. Under the statute, delisting would have required a finding that no EGU's emissions exceeded a level adequate to protect public health with an ample margin of safety, and that no adverse environmental effect would result from any source.

[13] For a link to the proposed rule as well as explanatory material, see U.S. EPA, "Reducing Toxic Air Emissions from Power Plants," at http://www.epa.gov/airquality/powerplanttoxics/ actions.html.

[14] U.S. EPA, "Benefits and Costs of Cleaning Up Toxic Air Pollution from Power Plants," p. 2, at http://www.epa.gov/airquality/powerplanttoxics/pdfs/20111221MATSimpactsfs.pdf.

[15] See U.S. EPA, "Memorandum: Emissions Overview: Hazardous Air Pollutants in Support of the Final Mercury and Air Toxics Standard," November 2011, Tables 5 and 6, available at http://www.epa.gov/airquality/powerplanttoxics/pdfs/20111216EmissionsOverviewMemo.pdf.

[16] EPA, Regulatory Impact Analysis, p. 3-3.

[17] Ibid., pp. 3-30 and 3-19.

[18] U.S. Department of Energy, Energy Information Administration, at http://www.eia.gov/electricity/sales_revenue_price/pdf/table3.pdf.

[19] The estimated price increase is from EPA, Regulatory Impact Analysis, p, 3-24. According to U.S. DOE's Energy Information Administration, the average monthly residential electric bill was $103.67 in 2009. See http://205.254.135.7/cneaf/electricity/esr/table5.html.

[20] EPA, Regulatory Impact Analysis, pp. 3-24 and 3-25.

[21] Scrubbers (more formally, flue gas desulfurization equipment) generally operate by spraying a lime or limestone slurry into the flue gas stream. The slurry reacts with the flue gas, absorbing and neutralizing the pollutants, which are then removed by a fabric filter (also known as a "bag house") or an electrostatic precipitator, which collects particles by imparting an electrical charge to them. For more information, see U.S. EPA, "Air Pollution Control Technology Fact Sheets," at http://www.epa.gov/ttn/catc/dir1/ffdg.pdf and at http://www.epa.gov/ttncatc1/dir1/fdespwpi.pdf.

[22] The data on projected installation and upgrade of pollution control equipment due to the rule are from EPA's Regulatory Impact Analysis, pp. 3-14 to 3-16.

[23] U.S. EPA, "Benefits and Costs of Cleaning Up Toxic Air Pollution from Power Plants," p. 1.

[24] For a discussion of major EPA rules recently proposed or promulgated, see CRS Report R41561, *EPA Regulations: Too Much, Too Little, or On Track?*

[25] EPA, Regulatory Impact Analysis, p. 3-16.

[26] EPA, Regulatory Impact Analysis, Chapter 6.

[27] Summary of Testimony, Michael J. Bradley, House Energy and Commerce Committee, "The American Energy Initiative: Recent EPA Rulemakings Relating to Boilers, Cement Manufacturing Plants, and Utilities," Hearing, April 15, 2011.

[28] ICF International, Potential Impacts of Environmental Regulation on the U.S. Generation Fleet, Final Report, prepared for the Edison Electric Institute, January 2011, p. 43, available at http://www.pacificorp.com/content/dam/pacificorp/doc/Energy_Sources/Integrated_ Resource_Plan/2011IRP/EEIModelingReportFinal-28January2011.pdf. Hereinafter, "EEI Report."

[29] North American Electric Reliability Corporation, *2010 Special Reliability Scenario Assessment: Resource Adequacy Impacts of Potential U.S. Environmental Regulations*, October 2010, http://www.nerc.com/files/EPA_Scenario_Final.pdf. NERC is an independent organization, founded by the electric utility industry, that conducts periodic, independent assessments of the reliability and adequacy of the bulk power system in North America.

[30] North American Electric Reliability Corporation, *2010 Special Reliability Scenario Assessment: Resource Adequacy Impacts of Potential U.S. Environmental Regulations*, October 2010, p. 50, http:// www.nerc.com/files/ EPA_Scenario_Final.pdf. Hereinafter referred to as the "2010 NERC report." NERC is an independent organization, founded by the electric utility industry, that conducts periodic, independent assessments of the reliability and adequacy of the bulk power system in North America.

[31] For a detailed comparison of equipment cost, see EEI report, p. 33.

[32] NERC, *2011 Long-Term Reliability Assessment*, November 2011, p. 173, at http:// www.nerc.com/files/ 2011LTRA_Final.pdf. Hereafter, "2011 NERC Report."

[33] "Derating" refers to the loss of available capacity because of the power needed to operate the pollution control equipment.

[34] 2011 NERC Report, pp. 197-486. The report also states in the text and in several figures that two regions in the Midwest will experience reserve margins below 15%, but these statements appear inconsistent with the detailed analyses of each region that begin on page 197.

[35] 2011 NERC Report, p. 156.

[36] 2011 NERC Report, p. 295.

[37] 2011 NERC Report, p. 156.

[38] For a map of NERC's Long Term Assessment Areas, see p. 4 of the 2011 NERC Report. The states include all or most of Alabama, Georgia, Illinois, Indiana, Iowa, Michigan, Minnesota, Missouri, New Jersey, Ohio, Pennsylvania, Virginia, West Virginia, and Wisconsin, and parts of Florida, Mississippi, North Carolina, and North Dakota.

[39] 2010 NERC Report, p. V.

[40] The White House, Office of the Press Secretary, "Presidential Memorandum—Flexible Implementation of the Mercury and Air Toxics Standards Rule," at http://www. whitehouse.gov/the-press-office/2011/12/21/presidential-memorandum-flexible-implementation-mercury-and-air-toxics-s.

[41] EPA, Mercury and Air Toxics Standards, pre-publication copy, pp. 576-591, at http:// www.epa.gov/airquality/powerplanttoxics/pdfs/20111216MATSfinal.pdf.

[42] "The Environmental Protection Agency's Enforcement Response Policy for Use of Clean Air Act Section 113(a)Administrative Orders in Relation to Electric Reliability and the Mercury and Air Toxics Standard," Memorandum from Cynthia Giles, Assistant Administrator, Office of Enforcement and Compliance Assurance, to EPA Regional Administrators et al., December 16, 2011, 7 p. at http://www.epa.gov/airquality/powerplanttoxics/pdfs/ EnforcementResponsePolicyforCAA113.pdf.

[43] Ibid., p. 4.

In: The Transformation of Electrical Power: Key Issues
Editor: Dawson B. Munoz
ISBN: 978-1-62618-619-4

Chapter 7

NUCLEAR ENERGY: OVERVIEW OF CONGRESSIONAL ISSUES*

Mark Holt

SUMMARY

The policy debate over the role of nuclear power in the nation's energy mix is rooted in the technology's fundamental characteristics. Nuclear reactors can produce potentially vast amounts of energy with relatively low consumption of natural resources and emissions of greenhouse gases and other pollutants. However, facilities that produce nuclear fuel for civilian power reactors can also produce materials for nuclear weapons. The process of nuclear fission (splitting of atomic nuclei) to generate power also results in the production of radioactive material that must be contained in the reactor and can remain hazardous for thousands of years. How to manage the weapons proliferation and safety risks of nuclear power, or whether nuclear power is worth those risks, are issues that have long been debated in Congress.

The 104 licensed nuclear power reactors at 65 sites in the United States generate about 20% of the nation's electricity. Five new reactors are currently licensed for construction. About a dozen more are planned, but whether they move forward will depend largely on their economic competitiveness with natural gas and coal plants. Throughout the world, 436 reactors are currently in service, and 62 more are under construction.

The March 2011 disaster at the Fukushima Dai-ichi nuclear power plant in Japan increased attention to nuclear safety throughout the world. The U.S. Nuclear Regulatory Commission (NRC), which issues and enforces nuclear safety requirements, established a task force to identify lessons from Fukushima applicable to U.S. reactors. The task force's report led to NRC's first Fukushima-related regulatory requirements on March 12, 2012. Several other countries, such as Germany and Japan, eliminated or reduced their planned future reliance on nuclear power after the accident.

Highly radioactive spent nuclear fuel that is regularly removed from nuclear power plants is currently stored at plant sites in the United States. Plans for a permanent underground repository at Yucca Mountain, NV, were abandoned by the Obama

* This is an edited, reformatted and augmented version of the Congressional Research Service Publication, CRS Report for Congress R42853, dated December 11, 2012.

Administration, although that decision is being challenged in court. The Obama Administration appointed the Blue Ribbon Commission on America's Nuclear Future to recommend an alternative nuclear waste policy. The Commission recommended in January 2012 that new candidate sites for nuclear waste storage and disposal facilities be selected through a "consent based" process.

The level of security that must be provided at nuclear power plants has been a high-profile issue since the 9/11 terrorist attacks on the United States in 2001. Since those attacks, NRC issued a series of orders and regulations that substantially increased nuclear plant security requirements, although industry critics contend that those measures are still insufficient.

Encouraging exports of U.S. civilian nuclear products, services, and technology while making sure they are not used for foreign nuclear weapons programs has long been a fundamental goal of U.S. nuclear energy policy. Recent proposals to build nuclear power plants in several countries in the less developed world, including the Middle East, have prompted concerns that international controls may prove inadequate.

SYNTHESIS OF KEY ISSUES

The long-running public policy debate over the future of nuclear energy is rooted in the technology's inherent characteristics. Initially developed for its unprecedented destructive power during World War II, nuclear energy seemed to hold equal promise after the war as a way of providing limitless energy to all mankind. International diplomacy has focused ever since on finding institutional mechanisms for spreading the perceived benefits of nuclear energy throughout the world while preventing the technology from being used for the proliferation of nuclear weapons. Much of this international effort is focused on key nuclear fuel cycle facilities—plants for enriching uranium in the fissile isotope U-235 and for separating plutonium from irradiated nuclear fuel. Such plants can be used to produce civilian nuclear reactor fuel as well as fissile material for nuclear warheads.

Yet even the use of nuclear power solely for peaceful energy production has proven intrinsically controversial. The harnessing of nuclear fission in a reactor creates highly radioactive materials that must be kept from overheating and escaping from the reactor building, as occurred during the disasters at Fukushima and Chernobyl. Spent nuclear fuel that is regularly removed from reactors during refueling must be isolated from the environment for up to a million years. Potential technologies to reduce nuclear waste through recycling usually involve separating plutonium that could be used for nuclear weapons and would still leave substantial amounts of radioactive waste to be stored and disposed of. Long-term storage and disposal sites for nuclear waste have proven difficult to develop throughout the world, as illustrated by the Obama Administration's cancellation of the proposed U.S. waste repository at Yucca Mountain, NV.

The March 2011 disaster at Japan's Fukushima Dai-ichi nuclear power plant, which forced the evacuation of areas as far as 30 miles away, has slowed nuclear power expansion plans around the world, particularly in Japan and Western Europe. However, dozens of new reactors are still being planned and built in China, India, Eastern Europe, and elsewhere.[1] In these areas, nuclear power's initial promise of generating large amounts of electricity without the need for often-imported fossil fuels, along with the more recent desire to reduce greenhouse gas emissions, remains a compelling motivation.

With 104 licensed reactors, the United States has the largest nuclear power industry in the world. But U.S. nuclear power growth has been largely stagnant for the past two decades, as natural gas has captured most of the market for new electric generating capacity.[2] Congress enacted incentives for new nuclear plants in the Energy Policy Act of 2005 (P.L. 109-58), including production tax credits, loan guarantees, and insurance against regulatory delays. Those incentives, combined with rising natural gas prices and concerns about federal restrictions on carbon dioxide emissions, prompted industry plans by late 2009 for up to 30 new nuclear power reactors in the United States.[3] However, falling natural gas prices and the defeat of greenhouse gas legislation in the 111th Congress have put many of those projects on hold. Currently, four reactors, in Georgia and South Carolina, have been licensed for construction and operation, and an older reactor on which construction had been suspended for two decades is now being completed in Tennessee. A variety of incentives to renew the growth of nuclear power have been proposed, including a plan by President Obama to include nuclear power, along with natural gas and advanced coal technologies, in a federal mandate for the production of "clean energy."

The extent to which the growth of nuclear power should be encouraged in the United States and around the world will continue to be a major component of the U.S. energy policy debate. Questions for Congress will include the implementation of policies to encourage or discourage nuclear power, post-Fukushima safety standards, development of new nuclear power and fuel cycle technologies, and nuclear waste management strategies.

BASIC FACTS AND STATISTICS

The 104 licensed nuclear power reactors at 65 sites in the United States generate about 20% of the nation's electricity. The oldest of today's operating reactors were licensed in 1969, and the most recent was in 1996. The reactors were initially licensed to operate for 40 years, but 80% have received or applied for 20-year license renewals by the Nuclear Regulatory Commission (NRC).Under the current mixture of 40- and 60-year licenses, 36 reactors would have to shut down by 2030 and the rest by 2049.[4]

Whether new reactors will be constructed to replace the existing fleet or even to expand nuclear power's market share will depend largely on costs. The cost of building and operating a new nuclear power plant in the United States is generally estimated to be significantly higher than natural gas combined-cycle plants (which use both combustion and steam turbines) and somewhat above conventional coal-fired plants. For example, the Energy Information Administration estimates that electricity generation from a nuclear power plant coming on line in 2016 would be 11.4 cents per kilowatt-hour (kwh), while combined-cycle gas would be 6.6 cents/kwh, conventional coal 9.5 cents/kwh, onshore wind 9.7 cents/kwh, offshore wind 24.3 cent/kwh, and solar photovoltaic 21.1 cents/kwh.[5] Such estimates depend on a wide range of variables, however, such as future fuel costs, environmental regulations, and tax credits and other incentives.

As noted above, the United States currently has five new reactors licensed for construction. They are scheduled to begin operating in 2015 through 2018.[6] Licenses to build and operate 14 additional reactors are currently pending at NRC, although their review schedules are uncertain.[7] If those additional U.S. reactors are licensed, they could begin coming on line in the early 2020s.

Throughout the world, 436 reactors are currently in service, and 62 more are under construction. France is the most heavily nuclear-reliant country in the world, with 58 reactors generating 78% of the country's electricity in 2011. Thirty countries in 2011 generated at least some of their electricity from nuclear power. After the Fukushima accident, Germany, which had previously generated about 30% of its electricity with nuclear power, closed eight of the country's 17 power reactors and decided to shut the remainder by 2022.[8] Japan, which had also generated about 30% of its electricity with nuclear power and had planned to raise that level to 50%, is reconsidering its energy policy. Only two Japanese reactors are currently operating, with the remaining 48 operable reactors undergoing safety improvements and regulatory reviews.

MAJOR NUCLEAR ENERGY ISSUES

Safety

The Fukushima Dai-ichi disaster, triggered by a huge earthquake and tsunami, greatly increased concerns about safety in the nuclear policy debate. The accident clearly demonstrated the potential consequences of a total loss of power (or "station blackout") at today's commercial nuclear plants. Even when a reactor shuts down, as the Fukushima plant did after the initial earthquake, residual radioactivity in the reactor core continues to generate heat that must be removed, typically by electrically driven cooling systems. When the tsunami knocked out power at three of the Fukushima reactors, the buildup of heat and pressure became so great that it melted the reactors' nuclear fuel and exceeded the limits of their containment structures. Cooling was also lost in Fukushima's spent fuel storage pools, causing concern that they could overheat, although later examination indicated that they did not.

Safety requirements for nuclear power plants are established and enforced in the United States by NRC, an independent regulatory commission. NRC safety regulations address the effects of external events such as earthquakes and floods, equipment failure such as breaks in coolant pipes, and other problems that could lead to radioactive releases into the environment. Critics of nuclear power contend that NRC is often reluctant to impose necessary safety requirements that would be costly or disruptive to the nuclear industry. However, the industry has frequently contended that costly safety proposals are unnecessary and would not significantly increase large existing safety margins.

Recent Events

Following the Fukushima disaster, NRC established a task force to identify lessons applicable to U.S. reactors and recommend safety improvements. The task force's report led to NRC's first Fukushima-related regulatory requirements on March 12, 2012. NRC ordered all reactors to develop strategies to maintain cooling and containment integrity during external events, such as floods and earthquakes, that were more severe than anticipated by the plants' designs ("beyond design basis"). In addition, NRC required that U.S. reactors of similar design to the Fukushima reactors have "reliable hardened vents" to remove excess pressure from their primary containments, and that better instrumentation be installed to monitor the condition of spent fuel pools during accidents.[9] NRC staff announced on November 1, 2012, a proposal to require the newly mandated containment vents to include filters to reduce the release of radioactive material if the vents have to be used. The idea drew praise from nuclear critics but opposition from the industry on cost grounds.[10]

Selected Congressional Action

Nuclear Power Plant Safety Act of 2011 (H.R. 1242, Markey)

Requires NRC to revise its regulations within 18 months to ensure that nuclear plants could handle major disruptive events, a loss of off-site power for 14 days, and the loss of diesel generators for 72 hours. Spent fuel would routinely have to be moved from pool to dry-cask storage within a year after it had cooled sufficiently, and emergency planning would have to include multiple concurrent disasters. NRC could not issue new licenses or permits until the revised regulations were in place. Introduced March 29, 2011; referred to Committee on Energy and Commerce.

Oversight Hearing: NRC's Implementation of Recommendations for Enhancing Nuclear Reactor Safety in the 21st Century

Joint hearing before the Senate Committee on Environment and Public Works and Subcommittee on Clean Air and Nuclear Safety. September 12, 2012. Witnesses: NRC Chairman Allison M. Macfarlane and NRC commissioners.

CRS Reports

CRS Report RL33558, *Nuclear Energy Policy*, by Mark Holt

CRS Report R41805, *Nuclear Power Plant Design and Seismic Safety Considerations*, by Anthony Andrews and Peter Folger

CRS Report R41694, *Fukushima Nuclear Disaster*, by Mark Holt, Richard J. Campbell, and Mary Beth Nikitin

Additional References

Recommendations for Enhancing Reactor Safety in the 21st Century, Nuclear Regulatory Commission, Near-Term Task Force Review of Insights from the Fukushima Dai-ichi Accident, July 12, 2011, http://pbadupws.nrc.gov/ docs/ML1118/ML111861807.pdf.

State-of-the-Art Reactor Consequence Analyses (SOARCA) Report: Draft Report for Comment, Nuclear Regulatory Commission, NUREG-1935, January 2012, http://www.nrc.gov/reading-rm/ doc-collections/nuregs/ staff/sr1935.

Radioactive Waste

Highly radioactive spent nuclear fuel must regularly be removed from operating reactors and stored in adjacent pools of water. After several years of cooling, the spent fuel can be placed in dry casks for storage elsewhere on the plant site. When the reactors were built, it had been expected that spent fuel would be taken away for reprocessing (separation of plutonium and uranium to make new fuel) or permanent disposal. However, reprocessing has not become commercialized in the United States, for economic and nonproliferation reasons, and central waste storage and disposal facilities have proven difficult to site. As a result, the vast majority of U.S. commercial spent fuel remains at the nuclear plants where it was generated—currently totaling 67,450 metric tons and rising at the rate of about 2,000 metric tons per year.[11]

Recent Events

The Nuclear Waste Policy Act (P.L. 97-425, NWPA), as amended in 1987, named Yucca Mountain, NV, as the nation's sole candidate site for a permanent high-level nuclear waste repository. However, the Obama Administration decided to halt the Yucca Mountain project and appointed the Blue Ribbon Commission on America's Nuclear Future to recommend an alternative policy. The Commission recommended in January 2012 that new candidate sites for nuclear waste storage and disposal facilities be found through a "consent based" process. The Department of Energy (DOE) is currently preparing legislation to implement the Commission's recommendations. At the same time, lawsuits are pending in federal court to require the development of the Yucca Mountain repository to continue, as specified by NWPA.

Selected Congressional Action

Restore America Act of 2011 (H.R. 3302, Rooney)

Among other provisions, encourages tripling of U.S. nuclear power capacity, requires licensing proceedings to continue for the proposed Yucca Mountain waste repository, removes statutory capacity limits on the repository, prohibits the President from blocking or hindering nuclear spent fuel recycling, establishes a nuclear fuel reserve, and establishes expedited reactor licensing procedures. Introduced November 1, 2011; referred to multiple committees.

Yucca Utilization to Control Contamination Act (H.R. 4625, Joe Wilson)/Nuclear Waste Fund Relief and Rebate Act (S. 2176, Graham)

Requires that payments into the Nuclear Waste Fund be returned to utilities unless the President certifies that Yucca Mountain is the selected site for a nuclear waste repository; that defense nuclear waste be transported to Yucca Mountain beginning in 2017; and that statutory requirements for disposal of nuclear waste be sufficient grounds for NRC to determine that waste from new or relicensed reactors will be disposed of in a timely manner. House bill introduced April 25, 2012; referred to Committee on Energy and Commerce. Senate bill introduced March 8, 2012; referred to Committee on Energy and Natural Resources.

Nuclear Fuel Storage Improvement Act of 2011 (S. 1320, Murkowski)

Authorizes the Secretary of Energy to provide payments to units of local government that, with the approval of the state governor, volunteer to host a "privately owned and operated temporary used fuel storage facility." Introduced June 30, 2011; referred to Committee on Environment and Public Works.

Nuclear Waste Administration Act of 2012 (S. 3469, Bingaman)

Establishes the Nuclear Waste Administration to develop central storage and disposal facilities for high-level nuclear waste. Sites would require the consent of the host state and local governments and Indian tribes. Introduced August 1, 2012; referred to Committee on Energy and Natural Resources. Committee hearing held September 12, 2012.

Oversight Hearing: Recommendations of the Blue Ribbon Commission on America's Nuclear Future

Hearing before the Subcommittee on Environment and the Economy of the House Committee on Energy and Commerce. February 1, 2012. Lead witnesses: Hon. Lee Hamilton and Lt. Gen. Brent Brent Scowcroft (Ret.), co-chairs of the Blue Ribbon Commission on America's Nuclear Future.

CRS Reports

CRS Report RL33461, *Civilian Nuclear Waste Disposal*, by Mark Holt

CRS Report R42513, *U.S. Spent Nuclear Fuel Storage*, by James D. Werner

CRS Report R40996, *Contract Liability Arising from the Nuclear Waste Policy Act (NWPA) of 1982*, by Todd Garvey

Additional References

Report to the Secretary of Energy, Blue Ribbon Commission on America's Nuclear Future, January 2012, http://cybercemetery.unt.edu/archive/brc/ 20120620211605/http:/brc.gov.

Managing Spent Nuclear Fuel: Strategy Alternatives and Policy Implications, RAND Corporation, 2010, http://www.rand.org/pubs/monographs/ MG970.html.

Energy and Environmental Policy

Congress has long debated the role that nuclear power should play in meeting national energy and environmental goals. Nuclear power supporters generally point to the technology as crucial for providing a secure, domestic source of energy with low emissions of greenhouse gases and other emissions. Opponents generally counter that safety and proliferation risks, nuclear waste hazards, and high costs outweigh those benefits. The debate over nuclear power's role often focuses on the level of federal support that should be provided to encourage the construction of new nuclear plants, through such mechanisms as loan guarantees, tax credits, clean energy mandates, and liability limits. Because of the relatively high cost of new nuclear reactors, especially compared with natural gas plants, the level of federal support is expected to be a key determinant of the future growth or decline of nuclear power in the United States. Federal support for nuclear energy research and development could also be an important factor in the long term.

Recent Events

One nuclear power project, consisting of two new reactors at the Vogtle plant in Georgia, received a conditional commitment from DOE for an $8.33 billion loan guarantee in February 2010, as authorized by Section 1703 of the Energy Policy Act of 2005 (P.L. 109-58). However, negotiations over the detailed terms of the loan guarantee have continued since then, particularly over the level of fees that the borrower would have to pay, so it has not been finalized. No other planned nuclear plants have received conditional commitments for DOE loan guarantees.

Selected Congressional Action

Clean Energy Financing Act of 2011 (S. 1510, Bingaman)

Establishes a Clean Energy Deployment Administration to provide financial assistance to commercial projects using clean energy technology, including nuclear power. Introduced and reported as an original measure by the Committee on Energy and Natural Resources August 30, 2011 (S.Rept. 112-47).

Clean Energy Standard Act of 2012 (S. 2146)

Establishes minimum U.S. annual percentages of clean energy use, including nuclear power, starting at 24% in 2015 and rising to 84% in 2035. Introduced March 1, 2012; referred to Committee on Energy and Natural Resources. Committee hearing held May 17, 2012.

CRS Reports

CRS Report RL33558, *Nuclear Energy Policy*, by Mark Holt

Additional References

Maintaining U.S. Leadership in Global Nuclear Energy Markets, Bipartisan Policy Center, September 2012, http://bipartisanpolicy.org/sites/ default/files/Nuclear%20Report.PDF.

World Nuclear Industry Status Report 2012, Mycle Scheider Consulting, July 2012, http://www.worldnuclearreport.org/IMG/pdf/2012MSC-WorldNuclearReport-EN-V2.pdf.

Security and Emergency Response

The level of security that must be provided at nuclear power plants has been a high-profile issue since the 9/11 terrorist attacks on the United States in 2001. Since those attacks, NRC issued a series of orders and regulations that substantially increased nuclear plant security requirements, although industry critics contend that those measures are still insufficient. Key measures include an increase in the level of attacks that nuclear plant security forces must be able to repel, requirements for mitigating the effects of large fires and explosions, and a requirement that new reactors be capable of withstanding aircraft crashes without releasing radioactive material. NRC also modified its planning requirements for evacuations and other emergency responses after the 9/11 attacks, and the Fukushima disaster illustrated the importance of emergency response to radioactive releases from any cause.

Recent Events

NRC issued wide-ranging revisions to its emergency preparedness regulations on November 1, 2011, dealing with duties of emergency personnel and the inclusion of hostile actions in emergency planning drills.[12] In response to Fukushima, NRC staff recommended that nuclear emergency plans be required to address events affecting multiple reactors and prolonged station blackout. NRC told nuclear power plants on March 12, 2012, to provide specific information and analysis on those issues.[13]

Selected Congressional Action

Nuclear Power Licensing Reform Act of 2011 (H.R. 1268, Lowey)

Requires evacuation planning within 50 miles of U.S. nuclear power plants and that reactor license renewals be subject to the same standards that would apply to new reactors. Introduced April 7, 2011; referred Committee on Energy and Commerce.

Nuclear Disaster Preparedness Act (H.R. 1694, Engle)

Requires the President to issue guidance for federal response to nuclear disasters, covering specific topics listed in the bill. Introduced May 3, 2011; referred to Committee on Transportation and Infrastructure.

CRS Reports

CRS Report RL34331, *Nuclear Power Plant Security and Vulnerabilities*, by Mark Holt and Anthony Andrews

Additional References

Backgrounder—Nuclear Security, Nuclear Regulatory Commission, October 2008 (last reviewed or updated October 3, 2012), http://www.nrc.gov/reading-rm/doc-collections/fact-sheets/security-enhancements.html.

Nuclear Weapons Nonproliferation

Encouraging exports of U.S. civilian nuclear products, services, and technology while making sure they are not used for foreign nuclear weapons programs has long been a fundamental goal of U.S. nuclear energy policy. International controls and inspections are intended to ensure the peaceful use of civilian nuclear facilities and prevent the proliferation of nuclear weapons. However, recent proposals to build nuclear power plants in as many as 18 countries[14] that have not previously used nuclear energy, including several in the Middle East and elsewhere in the less developed world, have prompted concerns that international controls may prove inadequate. Numerous recommendations have been made in the United States and elsewhere to create new incentives for nations to forgo the development of uranium enrichment and spent nuclear fuel reprocessing facilities that could produce weapons materials as well as civilian nuclear fuel.

Recent Events

Iran is currently the prime example of the tension between peaceful and weapons uses of nuclear technology. Of particular concern is a growing Iranian uranium enrichment program, which Iran contends is solely for peaceful purposes but which the United States and other countries suspect is for producing weapons material. The U.N. Security Council has imposed sanctions and passed several resolutions calling on Iran to suspend its enrichment program and other sensitive nuclear activities. Nevertheless, Iran continues to advance its nuclear program.

Extension of the U.S.-South Korea nuclear cooperation agreement, which expires in 2014, is also being affected by nonproliferation issues. South Korea would like to include advance U.S. consent for spent fuel reprocessing and uranium enrichment, but the United States is concerned about the precedent that such an agreement might set and how it would affect other ongoing issues on the Korean peninsula. Official talks on the extension began in October 2010, with two more rounds held since, most recently in February 2012.

Selected Congressional Action

To amend the Atomic Energy Act of 1954 to require congressional approval of agreements for peaceful nuclear cooperation with foreign countries, and for other purposes (H.R. 1280, Ros-Lehtinen/S. 109, Ensign)

Requires congressional approval of U.S. peaceful nuclear cooperation agreements with countries that do not agree to forgo enrichment and reprocessing. House bill introduced March 31, 2011; reported by Committee on Foreign Affairs May 30, 2012 (H.Rept. 112-507). Senate bill introduced January 25, 2011; referred to Committee on Foreign Relations.

CRS Reports

CRS Report RL34234, *Managing the Nuclear Fuel Cycle: Policy Implications of Expanding Global Access to Nuclear Power*, coordinated by Mary Beth Nikitin

CRS Report R41910, *Nuclear Energy Cooperation with Foreign Countries: Issues for Congress*, by Paul K. Kerr, Mark Holt, and Mary Beth Nikitin

CRS Report RL31559, *Proliferation Control Regimes: Background and Status*, coordinated by Mary Beth Nikitin

Other References

U.S. Nuclear Cooperation as Nonproliferation: Reforms, or the Devil You Know? Nuclear Threat Initiative, November 27, 2012, http://www.nti.org/ analysis/articles/us-nuclear-cooperation-nonproliferation-reforms-or-devil-you-know.

End Notes

[1] World Nuclear Association, "World Nuclear Power Reactors & Uranium Requirements," November 2012, http://www.world-nuclear.org/info/reactors.html.

[2] Energy Information Administration, "Most Electric Generating Capacity Additions in the Last Decade Were Natural Gas-Fired," July 5, 2011, http://www.eia.gov/todayinenergy/ detail.cfm?id=2070.

[3] Nuclear Regulatory Commission, "Expected New Nuclear Power Plant Applications," updated September 28, 2009. Available from the author.

[4] Nuclear Regulatory Commission, *Information Digest, 2012-2013*, NUREG-1350, Vol. 24, August 2012, Section 3: Nuclear Reactors, http://www.nrc.gov/reading-rm/doc-collections/nuregs/staff/sr1350.

[5] Energy Information Administration, "Levelized Cost of New Generation Resources in the Annual Energy Outlook2011," November 2010, http://www.eia.gov/oiaf/aeo/ electricity_generation.html.

[6] Tennessee Valley Authority, "Watts Bar Unit 2 Project Construction Update," October 26, 2012, http://www.tva.com/power/nuclear/wattsbar_unit2.htm; South Carolina Electric and Gas, "Project Schedule,"

http://www.sceg.com/en/about-sceg/power-plants/new-nuclear-development/schedule/default.htm; Southern Company, "Milestones," http://www. southerncompany.com/nuclearenergy/milestones.aspx.

[7] Nuclear Regulatory Commission, "New Reactor Licensing Applications," September 10, 2012, http://www.nrc.gov/reactors/new-reactors.html.

[8] World Nuclear Association, "Public Information Service," http://www.world-nuclear.org.

[9] Nuclear Regulatory Commission, "Actions in Response to the Japan Nuclear Accident: March 12, 2012," updated May 30, 2012, http://www.nrc.gov/reactors/operating/ops-experience/ japan/timeline/03122012.html.

[10] Freebairn, William, "NRC Staff Recommends Ordering Filtered Vents for 31 Power Reactors," *Inside NRC*, November 5, 2012, p. 1.

[11] Gutherman Technical Services, "2010 Used Fuel Data," Report to Nuclear Energy Institute, January 18, 2010.

[12] NRC, "Enhancements to Emergency Preparedness Regulations," final rule, *Federal Register*, November 23, 2011, p. 72560.

[13] NRC, "Request for Information Pursuant to Title 10 of the Code of Federal Regulations 50.54(f) Regarding Recommendations 2.1, 2.3, and 9.3 of the Near-Term Task Force Review of Insights from the Fukushima Dai-ichi Accident," March 12, 2012, http:// pbadupws.nrc.gov/docs/ML1205/ML12053A340.pdf.

[14] World Nuclear Association, "World Nuclear Power Reactors & Uranium Requirements," November 2012, http://www.world-nuclear.org/info/reactors.html.

In: The Transformation of Electrical Power: Key Issues
Editor: Dawson B. Munoz
ISBN: 978-1-62618-619-4

Chapter 8

FUEL USED IN ELECTRICITY GENERATION IS PROJECTED TO SHIFT OVER THE NEXT 25 YEARS*

U.S. Energy Information Administration

While coal is projected to retain the largest share of the electricity generation mix through 2035, analyses included in the Annual Energy Outlook 2012 (AEO2012) anticipate its share declining as more generation comes from natural gas and renewable technologies. Coal's role as the preeminent source of electricity generation in the United States has lessened in recent years, declining from 49% of total electricity generation in 2007 to 42% in 2011.

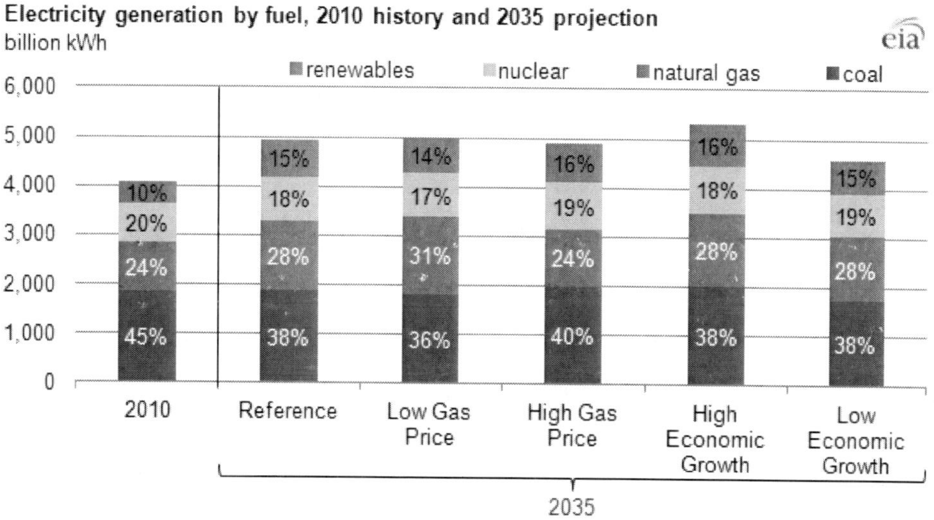

Source: U.S. Energy Information Administration, *Annual Energy Outlook 2012 Electric Power Projections*.

The shifts are due to changes in the comparative costs of electricity generation that result from changes in natural gas prices, coal prices, economic growth, and the implementation of

* This is an edited, reformatted and augmented version of the U.S. Energy Information Administration, dated July 30, 2012.

the *Cross-State Air Pollution Rule* and the *Mercury and Air Toxics Standards*. The Annual Energy Outlook includes alternative scenarios where the different assumptions about future fuel prices and economic growth illustrate the sensitivity of the electric power markets to these factors. For comparison with the projected values for 2035 in the chart above, the chart below presents projected electricity generation by fuel in 2020.

Projected fuel prices and economic growth are key factors influencing the future electricity generation mix. The price of natural gas, coal's chief competitor, has dropped significantly in recent years due to the increase in domestic production of natural gas. The sensitivity of the electric power market to changes in natural gas prices is illustrated in the high and low gas price alternative cases. The high gas price case results in less generation from natural gas relative to the Reference case. This results in coal, nuclear, and renewables retaining a larger share of the generation mix. Conversely, lower gas prices result in more generation from natural gas, whose share grows relative to those of the other sources. Fuel prices affect overall generation by influencing how existing plants are dispatched and in some cases whether or not they are retired. They also play a key role in the investment decisions for potential new plants.

AEO2012 also features cases that assume high and low economic growth relative to the Reference case. In the high economic growth case, coal generation increases relative to the Reference case as coal plants that might otherwise have retired remain operational to keep up with growth in the demand. On the other hand, weak economic conditions lead to lower electric demand, which reduces the need for generation.

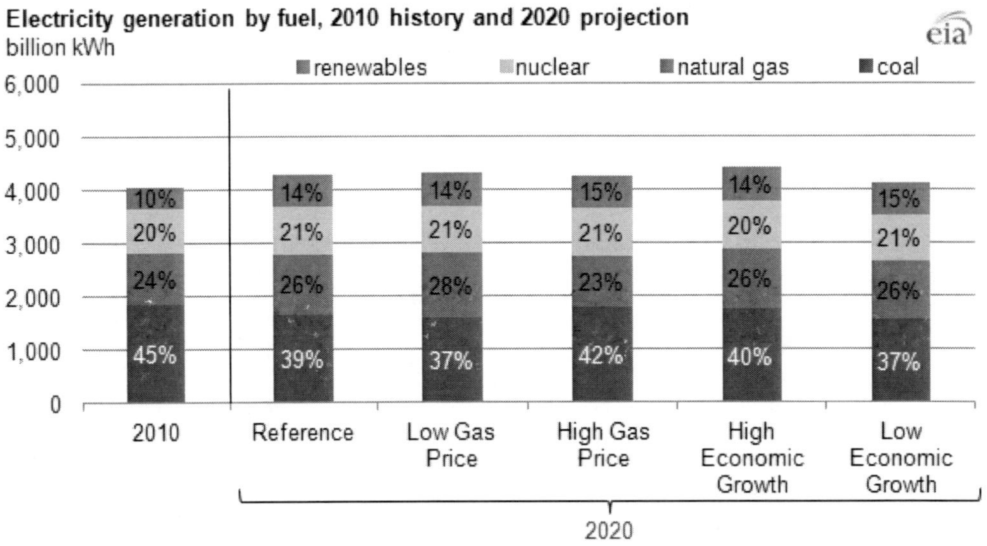

Source: U.S. Energy Information Administration, *Annual Energy Outlook 2012 Electric Power Projections*. (Original web link: http://www.eia.gov/todayinenergy/detail.cfm?id=7310)

In all cases highlighted, most of the growth in renewable electricity generation comes from wind and biomass facilities, which benefit from State Renewable Portfolio Standard (RPS) requirements, Federal tax credits, and, in the case of biomass, the availability of low-cost feedstocks and the *Renewable Fuel Standard (RFS)*. Overall, the amount of generation from nuclear power in 2035 increases from 2010 levels in all cases with the highest growth in the high gas price and high economic growth case. Despite the growth, nuclear's share of total generation in generation in 2035 is projected to fall below its 2010 level in all of the *cases*.

Every AEO2012 cases assume that all coal plants must have either a *Flue Gas Desulfurization (FGD) system or Direct Sorbent Injection system* installed by 2015 in order to continue operation in compliance with environmental rules. Currently, *over half of the coal plants in the nation* have FGD systems. Details relating to projected coal plant retirements in the AEO2012 analysis will be highlighted in an upcoming article.

INDEX

#

A

B

C

D

E

F

G

H

I

J

K

L

M

N

O

S

T

U

V

W

Y